Foundations of Earth Science

Foundations of Earth Science

Elliott Watts

STATES
ACADEMIC PRESS
www.statesacademicpress.com

States Academic Press,
109 South 5th Street,
Brooklyn, NY 11249, USA

Visit us on the World Wide Web at:
www.statesacademicpress.com

ISBN: 978-1-63989-216-7 (Hardback)

Cataloging-in-Publication Data

Foundations of earth science / Elliott Watts.
p. cm.
Includes bibliographical references and index.
ISBN 978-1-63989-216-7
1. Earth sciences. 2. Geology. 3. Physical sciences. 4. Environmental sciences.
I. Watts, Elliott.
QE26.3 .F68 2022
550--dc23

Table of Contents

Preface

The field of science which focuses on the physical and chemical composition of earth is known as earth science or geoscience. It is a sub branch of planetary science which studies the features of the lithosphere, atmosphere, hydrosphere and biosphere. Earth science makes use of physics, geography, chemistry, geology, chronology and mathematics to understand the evolution of earth. Various fields of study are covered under it are glaciology, soil sciences, hydrology, geochemistry, geophysics and atmospheric sciences. The principles of earth science are used for locating energy and mineral resources. Various earth processes such as earthquakes, hurricanes and volcanoes are covered by earth sciences for developing effective mitigation strategies. These studies rely on a combination of observational and experimental approaches. The book studies, analyses and uphold the pillars of earth science and its utmost significance in modern times. Different approaches, evaluations and methodologies and advanced studies on earth science have been included herein. Coherent flow of topics, student-friendly language and extensive use of examples make this book an invaluable source of knowledge.

A foreword of all chapters of the book is provided below:

Chapter 1- Earth science is the study of the physical and chemical constitution of the Earth and its atmosphere. It includes geology, hydrology and meteorology. It aims to gain a better understanding of the Earth's present and past evolution. This introductory chapter presents the subject matter and its various interesting and practical applications in a clear and concise manner.

Chapter 2- The Earth has undergone changes both externally and internally over millions of years. This chapter focuses on the fundamental processes that shaped and influenced the landscapes and topography of the Earth over the course of time. The different weathering processes and the resulting landforms are covered in detail in this chapter.

Chapter 3- Atmosphere is the layer of gases that surround the Earth it is responsible for the various natural phenomena that make it habitable. The atmosphere has evolved over the course of the Earth's existence in both content and structure. Its structure, evolution, divisions and associated processes have been covered extensively in this chapter to provide a better understanding of the subject matter.

Chapter 4- Oceans cover more than 70 percent of the surface of the Earth. The study of the different features of the ocean is known as oceanography. It includes physical, biological, chemical, geological and coastal oceanography. This chapter closely examines the key oceanography to provide an extensive understanding of the subject.

Chapter 5- Sedimentology is the scientific discipline that is concerned with the study of sedimentary rocks, their chemical and physical properties, formation, transportation and deposition. Sedimentary rocks record the events and processes that shaped the Earth's surface and their study is critical to understand the geological history of the Earth. This chapter deals with this discipline in a comprehensive manner.

Chapter 6- Meteorology is the study of the atmosphere and the various phenomena associated with it along with their effects on the weather. Climate and global weather conditions are influenced by the atmosphere and its study is known as climatology. This chapter closely examines both these area of study in order to provide a thorough understanding of the related topics.

Chapter 7 - Many sciences like geophysics, geochemistry, geobiology and geoinfomatics are branches of earth science. They study the various chemical, physical and biological phenomena associated with the Earth. The topics elaborated in this chapter will help in gaining a better perspective of all the related aspects of earth and planetary science and their applications.

At the end, I would like to thank all the people associated with this book devoting their precious time and providing their valuable contributions to this book. I would also like to express my gratitude to my fellow colleagues who encouraged me throughout the process.

Elliott Watts

Chapter 1

Earth Science: An Introduction

Earth science is the study of the physical and chemical constitution of the Earth and its atmosphere. It includes geology, hydrology and meteorology. It aims to gain a better understanding of the Earth's present and past evolution. This introductory chapter presents the subject matter and its various interesting and practical applications in a clear and concise manner.

Earth Science is the study of the Earth and its neighbors in space. It is an exciting science with many interesting and practical applications. Some Earth scientists use their knowledge of the Earth to locate and develop energy and mineral resources. Others study the impact of human activity on Earth's environment, and design methods to protect the planet. Some use their knowledge about Earth processes such as volcanoes, earthquakes, and hurricanes to plan communities that will not expose people to these dangerous events.

The Four Earth Sciences

Many different sciences are used to learn about the Earth; however, the four basic areas of Earth science study are: geology, meteorology, oceanography, and astronomy.

Geology: Science of the Earth

Geology is the primary Earth science. The word means "study of the Earth." Geology deals with the composition of Earth materials, Earth structures, and Earth processes. It is also concerned with the organisms of the planet and how the planet has changed over time. Geologists search for fuels and minerals, study natural hazards, and work to protect Earth's environment.

Meteorology: Science of the Atmosphere

Meteorology is the study of the atmosphere and how processes in the atmosphere determine Earth's weather and climate. Meteorology is a very practical science because everyone is concerned about the weather. How climate changes over time in response to the actions of people is a topic of urgent worldwide concern. The study of meteorology is of critical importance in protecting Earth's environment.

Oceanography: Science of the Oceans

Oceanography is the study of Earth's oceans - their composition, movement, organisms and processes. The oceans cover most of our planet and are important resources for food and other commodities. They are increasingly being used as an energy source. The oceans also have a major influence on the weather, and changes in the oceans can drive or moderate climate change. Oceanographers work to develop the ocean as a resource and protect it from human impact. The goal is to utilize the oceans while minimizing the effects of our actions.

Astronomy: Science of the Universe

Astronomy is the study of the universe. Here are some examples of why studying space beyond Earth is important: the moon drives the ocean's tidal system, asteroid impacts have repeatedly devastated Earth's inhabitants, and energy from the sun drives our weather and climates. Knowledge of astronomy is essential to understanding the Earth. Astronomers can also use knowledge of Earth materials, processes and history to understand other planets - even those outside of our own solar system.

The Importance of Earth Science

Today we live in a time when the Earth and its inhabitants face many challenges. Our climate is changing, and that change is being caused by human activity. Earth scientists recognized this problem and will play a key role in efforts to resolve it. We are also challenged to: develop new sources of energy that will have minimal impact on climate; locate new sources of metals and other mineral resources as known sources are depleted; and, determine how Earth's increasing population can live and avoid serious threats such as volcanic activity, earthquakes, landslides, floods and more. These are just a few of the problems where solutions depend upon a deep understanding of Earth science.

Earth science investigations are increasingly utilizing geospatial technologies. The rapid proliferation of geospatial technologies is related to advances in geodesy, photogrammetry, geophysics, computer science, statistics, remote sensing, geographic information technology (GIT), and numerical modeling, which have collectively revolutionized the field of geomorphology. Scientists are routinely utilizing new spatio-temporal data, geocomputational algorithms, and processing approaches and models, which now allow assessments far beyond traditional geomorphological mapping. It is now possible to quantify landscape morphology, assess surface biophysical conditions, assess near-surface conditions, link process with form and patterns and improve our understanding of scale dependence and the polygenetic nature of landscape evolution. Nevertheless, such rapid utilization of information technologies must be carefully examined, given the empirical nature of utilizing geographic information systems (GIS), the need for formalization to address numerous issues, and requirements of accuracy and repeatability.

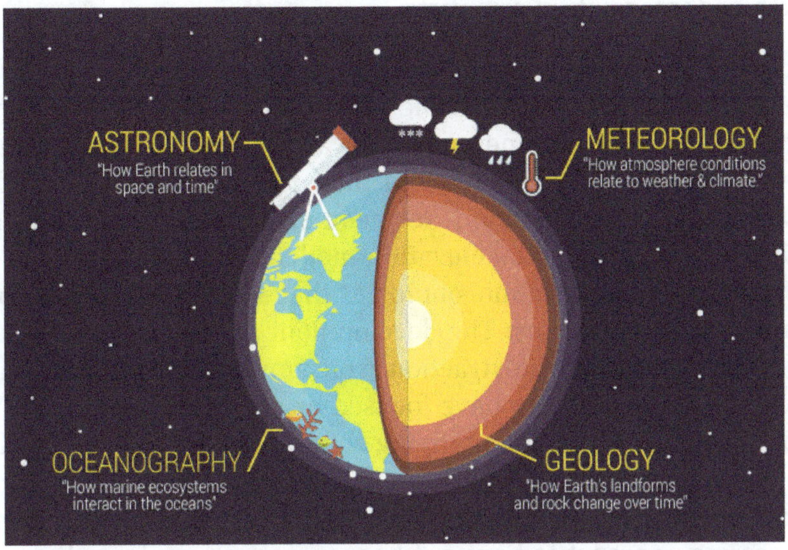

Numerous conceptual/theoretical and methodological issues are at the heart of effectively utilizing spatio-temporal data and GIT to study geomorphological systems. Therefore, Earth scientists need to be fully aware of current capabilities as well as the issues and challenges related to geomorphology and geographic information science (GIScience). Geospatial technologies can be used to address various conceptual and practical issues such as heterogeneous surface composition with fuzzy-classification membership, in determinant boundaries and features, hierarchical organization and spatial analysis using object-oriented technology, scale dependence of properties and patterns using geostatistics and objective mapping using different analytical approaches (e.g., descriptive statistics, inferential statistics, artificial intelligence, and various analytical reasoning technologies).

Nevertheless, numerous limitations are associated with the use of existing cartographic representations of environmental and geomorphological information, as parameterization schemes that uniquely characterize the formal structure of natural systems and information associations (i.e., attributes, space, and time), and the formal linkage of multiple processes and forms in space–time is not readily available. Consequently, cartographic representational schemes are still effectively being used, although more complex representational schemes typically occur outside of GIS environments. Addressing numerous science issues will most likely require multidisciplinary collaboration between Earth and information scientists.

From a more methodological perspective, advances in the development and deployment of new sensors and mobile platforms will provide for new forms of data across a multitude of scales. This has enabled researchers to address numerous issues at the requisite scale. Active research areas include the assessment of surface processes and their impact on erosion and landscape evolution and the evaluation of data and methods for geomorphological mapping. The role of remote sensing in generating high-quality digital elevation model (DEMs) is essential, as the topography inherently incorporates a multitude of morphologies that are generated due to the interaction of climate, tectonic, and surface processes. The development and evaluation of new techniques and analytical approaches for information extraction from multispectral, multitemporal, and DEMs is an active research area. Investigators have focused on the technical aspects of developing GIS databases, developing geomorphometric mapping software, mapping specific landform features, and developing new ways to visualize geomorphological information. A plethora of quantitative metrics and approaches exist. The advantages and limitations, however, of numerous algorithms and multistage processing approaches have not been rigorously evaluated and compared for specific applications. Furthermore, standardized formalizations for specific applications are urgently needed. Consequently, numerous issues related to taxonomy, scale, process mechanics, feedback mechanisms, system dynamics and states, representational schemes, algorithms and processing protocols, visualization of complex information, and effective information distribution need to be accounted for.

In Earth sciences, the assessment of geodiversity is often identified with the valuation and assignment of the advantages of abiotic components of the natural environment, as well as the appraisal of the relationships determining their dynamics and even the satisfaction of human needs. Therefore, the assessment of geodiversity usually takes the form of a classification. As Johnston noticed, by performing classification there are certain groups of entities or phenomena that can be treated as individual units, and, based on their specific patterns and behaviors it is possible to make significant generalisations. However, it should be borne in mind that classification alone does not necessarily contain any assessment and vice versa, not every assessment must include a classification.

The choice of a specific method to assess geodiversity and a proper selection of criteria for the assessment of geodiversity is dependent on many factors. The most important are: (1) the purpose of the assessment; (2) the type of landscape in the study area and its spatial dimension; (3) and above all, the availability of spatial data in an appropriate scale, including in its digital form.

Chapter 2

The Dynamic Earth

The Earth has undergone changes both externally and internally over millions of years. This chapter focuses on the fundamental processes that shaped and influenced the landscapes and topography of the Earth over the course of time. The different weathering processes and the resulting landforms are covered in detail in this chapter.

Weathering Processes

Weathering is the disintegration or alteration of rock in its natural or original position at or near the Earth's surface through physical, chemical, and biological processes induced or modified by wind, water, and climate.

Weathering: Weathered faces of the Pancake Rocks.

During the weathering process the translocation of disintegrated or altered material occurs within the immediate vicinity of the rock exposure, but the rock mass remains in situ. Weathering is distinguished from erosion by the fact that the latter usually includes the transportation of the disintegrated rock and soil away from the site of the degradation. A broader application of erosion, however, includes weathering as a component of the general denudation of all landforms along with wind action and fluvial, marine and glacial processes. The occurrence of weathering at or near the Earth's surface also distinguishes it from the physical and chemical alteration of rock through metamorphism, which usually takes place deep in the crust at much higher temperatures.

Weathering involves physical, chemical, and biological processes acting separately or, more often, together to achieve the disintegration and decay of rock material. Physical weathering causes the disintegration of rock by mechanical processes and therefore depends on the application of force. Disintegration involves the breakdown of rock into its constituent minerals or particles with no decay of any rock-forming minerals. The principal sources of physical weathering are thermal expansion and contraction of rock, pressure release upon rock by erosion of overlaying materials, the alternate freezing and thawing of water between cracks and fissures within rock, crystal growth within rock, and the growth of plants and living organisms in rock. Rock alteration usually involves chemical weathering in which the mineral composition of the rock is changed, reorganized, or redistributed. The rock minerals are exposed to solution, carbonation, hydration, and oxidation by circulating waters. These effects on the mineral decomposition are added to the effects of living organisms and plants as nutrient extraction to alter rock.

Several factors control the type of weathering and the rate at which rock weathers. The mineralogical composition of a rock will determine the rate of alteration or disintegration. The texture of the rock will affect the type of weathering that is most likely to occur. Fine-grain rock will usually be more susceptible to chemical alteration but less susceptible to physical disintegration. The pattern of joints, fractures, and fissures within rock may provide an avenue for water to penetrate. Thus, shattered and fractured rock masses are more likely to undergo weathering than are monolithic structures. Climate will also control the type and rate of weathering by affecting the likelihood of freeze–thaw cycles and chemical reactions. Chemical weathering is more likely to occur and to be more effective in humid tropical climates, and disintegration of rock from freeze–thaw cycles is more likely to take place and to be more effective in sub-Arctic climates.

Types of Weathering

Weathering is a general term for several processes. In general, weathering is of three types as:

- Physical weathering.
- Chemical weathering.
- Biological weathering.

The chemical and physical weathering often goes hand in hand. Physical weathering can occur due to temperature, pressure, frost, etc. Living organisms may contribute to mechanical weathering, as well as chemical weathering. Lichens and mosses that are growing on bare rock surfaces may create a more humid chemical microenvironment. The attachment of these organisms to the rock surface enhances the physical as well as chemical breakdown of the surface layer of the rocks. These are all very slow and silent processes.

Physical Weathering

Physical weathering is also called as mechanical weathering. Mechanical weathering is related to the physical breakup of rocks into small pieces and fragments. In physical weathering, there is no change in the chemistry of the parent rock. The physical characteristics of rocks also influence the process of physical weathering. The characteristics are differential composition, particle size, the hardness and degree of cementation. Some rock bodies are very resistant due to their strong physical properties. They may not be attacked by physical weathering processes. The structure of rocks is fully favourable to weathering activities. Presence of joints in rock bodies are typical examples.

Chemical Weathering

Chemical and structural characteristics of minerals present in rocks are also expected to play a significant role in weathering. Chemical weathering changes the composition of rocks, often transforming them when water interacts with their minerals to create various chemical reactions. Chemical weathering is a gradual and ongoing process as the mineralogy of the rock adjusts to the near surface environment. For minerals of given particle size, chemical and crystalline characteristics determine the ease of decomposition. (e.g.) gypsum – sparingly soluble in water, is dissolved and removed in solution form under high rainfall. Ferro magnesium minerals are more susceptible to chemical weathering than feldspar and quartz. Tightness of packing of ions in crystals is yet another property for encouraging mineral weathering processes. Less tightly packed minerals like olivine and biotitic are easily weathered as compared to tightly packed zircon and muscovite (resistant).

Topography and Climate

Topography is an important factor in relating rocks with the atmospheric pressure, temperature and water vapor. The climatic condition tends to control the kind and rate of weathering. Under conditions of low rainfall, there is a dominance of physical weathering which reduces the size and increases the surface area with little change in volume. The increase in moisture content encourages both chemical as well as mechanical changes. This also creates new minerals and soluble products. The rates of weathering are generally fastest in humid tropical regions as there is sufficient moisture and warmth to encourage chemical decomposition. The easily weatherable minerals disappear on account of intense chemical weathering and more resistant products (hydrous oxides of Fe and Al) tend to accumulate. Climate controls the dominant type of vegetation which in turn controls the biochemical reactions in soils and mineral weathering.

Rock Type

The rock type determines the resistance of the rock to the weathering processes that operate in that particular environment. Each rock type is composed of a particular set of minerals, which are joined together by crystallisation, chemical bonding or cementing. When the forces of plate tectonics move these rocks from the environment in which they formed and expose them to the atmosphere they begin to weather.

Rock Structure

Highly jointed or faulted rocks present many planes of weakness along which weathering agents (e.g. water) can penetrate into the rock mass.

Erosion

The dynamism and efficiency of erosion determines how rapidly any weathered material is removed, how frequently fresh rock is exposed to weathering, and if deeply weathered profiles are preserved.

Time

The duration of the period that the same type of weathering has been operating, uninterrupted by climatic change, earth movements, and other factors, determines the degree and depth to which the rocks have been weathered.

Physical Weathering Processes

Physical weathering happens especially in places where there is little soil and few plants grow, such as in mountain regions and hot deserts. The types of physical weathering include; abrasion, crystallization, root wedging, insolation weathering, human mining, animal activity, tumbling, compressional stress, crushing waves, tensional stress. Physical weathering is more effective in areas which have a little vegetation, a large diurnal range of temperature or, temperatures fluctuating around 0 degrees Celsius.

Abrasion

The primary process of physical weathering is abrasion. It is the process in which rock blocks are reduced in their sizes. Sand and other particles normally move above rock bodies. The mechanical scraping of a rock surface by friction between rocks and these moving particles is known as abrasion. Particles are transported by wind, glacier, waves, and gravity, running water or erosion. After friction, the moving particles dislodge the loose and weak debris from the sides of the rock. These particles can be dissolved in the water and taken away to some other location.

Abrasion by water, ice, and wind processes loaded with sediment can have tremendous cutting power. The intensity of abrasion depends on the hardness, concentration, velocity and mass of the moving particles. The major incidences responsible for physical weathering are expansion resulting from unloading, crystal growth, thermal expansion, organic activity and colloidal plucking.

Mechanisms of Physical Weathering

The major mechanisms of physical weathering include:

- Freezing and thawing.
- Root wedging.
- Heat Spalling.

Recurring hotness and coldness is an important controlling factor in weathering. One of the main causes of physical weathering is the formation of ice in cracks or cavities within rocks.

Freezing and Thawing

Block disintegration occurs when rocks split along joints. This can happen as a result of repeated cycles of freezing and thawing. Freezing and thawing are the first set of processes. First, water soaks into the cracks or cavities, of the rocks existing below the surface and gets sealed up. Then, if the temperature falls low enough during night times, the water freezes. When water freezes into ice, its volume gets increased by nine percent. As a result of this, it expands (thaws) in the cracks and may push hard enough to split the rock into minor fragments. Under specific circumstances, this expansion is able to displace and create fractures in most of the rocks. Not all volumetric expansion is caused by the pressure of the freezing water; it can be caused by stresses in water that remains unfrozen.

Frost Weathering

Frost weathering is a collective name given for several mechanical weathering processes induced by stresses created by the freezing of water into ice. The term includes a variety of processes such as frost shattering, frost wedging and cryofracturing. The process may act on a wide range of spatial and temporal scales, from minutes to years and from dislodging mineral grains to fracturing boulders. Frost weathering is mainly driven by the frequency and intensity of freeze-thaw cycles and the properties of the materials subject to weathering. It is most pronounced in high altitude and latitude areas.

Root Wedging

The second mechanism is the root wedging. Plant root, have a great role to play in weathering. Plants are effective agents of mechanical weathering. Roots can penetrate through the cracks of rocks to depths of several meters. As the roots grow, they exert a tremendous amount of pressure on the walls of the cracks. This breaks them into pieces. Root wedging is a major process of mountains containing Forests and natural vegetation.

Heat Spalling

The weathering due to fluctuation in temperature is termed as 'Thermal Weathering'. It is observed in almost all the climatic zones. It is more intense in regions characterized by sharp temperature fluctuations, dry air, absence or poorly developed vegetation cover.

Heat Spalling is a major process of mechanical weathering. Heat from forest fires will cause the outer surface layers of rock to expand quickly and break away in spalls. Most igneous and metamorphic rocks are polyminerallic i.e. composed of several minerals. These minerals have different coefficients of thermal expansion and it causes differential expansion of minerals, thus gives rise to minute internal fracturing.

Even in monominerallic-rocks, the liner co-efficient of expansion of mineral differs from one direction parallel to the crystallographic axis to the other. Thus, monominerallic-rocks also disintegrate due to temperature changes.

Rocks composed of different coloured minerals also undergo differential expansion. This is due to the fact that dark minerals are more strongly heated than the lighter ones. The difference in their volumetric expansion may also lead to the development of cracks and gradual disintegration of the rock. This process is also known as 'granular'-disintegration'.

Exfoliation

There are two more processes of physical weathering as exfoliation and spheroidal weathering. Exfoliation denotes the loss of outer layers of rock bodies as it weathers and detaches from the main mass. This leaves a dome like structure. Exfoliation is a form of mechanical weathering in which curved plates of rock are stripped from rock below. Exfoliation occurs particularly in hot dry desert climates, and on sheets of rock. Exfoliation domes are typical geomorphic features. In this reprocess the separation of successive thin shells, or spalls happen from massive rocks like granites or basalts. This is common in regions that have moderate rainfall. The thickness of individual sheet or plate may be from a few millimetres to a few metres.

Spheroidal Weathering

Spheroidal weathering is a form of chemical weathering that affects systematically jointed bedrock that results in the formation of concentric or spherical layers of highly decayed rock within weathered bedrock that is known as saprolite. When saprolite is exposed by physical erosion, these concentric layers peel (spall) off as concentric shells. Spheroidal weathering is a unique process. It is a kind of chemical weathering of systematically jointed, massive rocks, including granite, dolerite, basalt and sedimentary rocks such as silicified sandstone. It occurs as the result of the chemical alteration of such rocks along intersecting joints.

Chemical Weathering Processes

The second major type of weathering is the Chemical Weathering:

- It is the process of decomposition of Earth's surface materials.
- It is done by various agents and chemical reaction.

The reactions, here, are exothermic and result in the increase in volume of minerals that contribute to the physical disruption of rock.

Effectiveness of Chemical Weathering

The effectiveness of chemical weathering is directly related to the:

- Mineral stability in weathering.
- Amount of surface area exposed, which itself is exponentially related to the density of fractures in a rock.

Rate of Chemical Weathering

The rate of Chemical weathering depends on various factors like:

- Temperature.
- Amount of surface area.
- Availability of water or natural acid.

Impacts of Chemical Weathering

Most chemical weathering results in:

- An increase in bulk with resulting strains and stresses within the rocks.
- Formation of some lower density minerals.
- Reduction in particle size and increase in surface area of masses.
- Growth of more mobile minerals and formation of more stable minerals.

Processes of Chemical Weathering

The major processes involved in chemical weathering are:

- Solution.
- Oxidation.
- Hydration and Hydrolysis.
- Carbonation – Dissolution.

Solution

Solution is the process of dissolving mineral constituents by water or acid. Most minerals have low solubility in pure water, but rain contains carbonic acid, so that carbonate minerals dissolve readily in acidic solutions. Some substances present in the rocks are directly soluble in water. The soluble substances are removed by the continuous action of water and the rock no longer remains solid and form holes, rills or rough surface and ultimately fall into pieces or decomposes. The action is considerably increased when the water is acidified by the dissolution of organic and inorganic acids.

Hydration

Hydration involves two processes as hydration and hydrolysis. Hydration involves absorption of water. Chemical combination of water molecules with a particular substance or mineral leading to a change in structure.

Soil forming minerals in rocks do not contain any water and they undergo hydration when exposed to humid conditions. Rain, streams, and seawater dissolve minerals from rocks, causing the rocks to crumble. For example water dissolves the mineral feldspar from granite, leaving grains of quartz, a mineral that forms sand. Upon hydration there is swelling and increase in volume of minerals. The minerals lose their luster and become soft. It is one of the most common processes in nature and works with secondary minerals, such as aluminium oxide and iron oxide minerals and gypsum. Some of the typical mineral weathering reaction can easily illustrate these:

1. Conversion of haematite into Limonite is a typical example:

$$\underset{\text{(Hematite) (Red)}}{2Fe_2O_3 + 3HOH} \longrightarrow \underset{\text{(Limonite) (Yellow)}}{2Fe_2O_3 .3H_2O}$$

2. Conversion of bauxite into hydrous aluminum oxide is another example:

$$\underset{\text{(Bauxite)}}{Al_2O_3 + 3HOH} \longrightarrow \underset{\text{(Hyd. aluminium Oxide)}}{Al_2O_3 .3H_2O}$$

3. Conversion of anhydrite into gypsum is another example:

$$CaSO_4 + 2H_2O \longrightarrow CaSO_4.2H_2O$$

(Anhydrite) (Gypsum)

4. Conversion of olivine into serpentine in yet another example:

$$3(MgO.FeO.SiO_2) + 2H_2O \longrightarrow 3MgO._2SiO_2.2H_2O + SiO_2 + 3H_2O$$

(Olivine) (Serpentine)

Hydrolysis

Hydrolysis is also a process of chemical weathering. It is due to the dissociation of H2O into H+ and OH- ions which chemically combine with minerals and bring about changes, such as exchange, decomposition of crystalline structure and formation of new compounds. Water acts as a weak acid on silicate minerals. Reaction between mineral and water leads to the formation of a new mineral or dissolved material. Hydrolysis is the most important process in chemical weathering. It involves formation of hydroxyl ions and does not represent a chemical change. Hydrolysis of feldspar, for example, produces clay:

$$KAlSi_3O_8 + H_2O \longrightarrow HAlSi_3O_8 + KOH$$

(Orthoclase) (Aluminosilicic Acid + clay)

Oxidation

Atmosphere contains 21% of free oxygen. Whereas the oxygen content of air dissolved in water is 30-35%. These two forms of oxygen are the most active chemical agents for weathering.

Oxidation of Minerals and Rocks is an effective process in decoloration and decomposition of materials. In this process, the atmospheric oxygen combines with the metal ions of minerals to form oxides (or hydroxides). Sulphide minerals become unstable & gradually substituted by sulphates. The equations of Oxidation (attack by oxygen) are as follows:

$$2Fe_2SiO_4 \text{ (olivine)} + H_2O + O_2 = FeO.OH \text{ (goethite)} + \text{dissolved silica}$$

In the process of oxidation, the mineral Geothite dehydrates to form hematite, which is a very stable iron oxide. Geothite is yellowish color, while hematite is brick red in color. Thus, rocks rich in iron oxides tend to form mainly red soils.

Example: $FeS_2 + 8HCO_2 + 7.5\ O_2 = Fe_2O_3 + 4SO_4^{-2} + 4CO_2 + 4H_2O$

Pyrite oxidizes rapidly to form hematite in the presence of water and oxygen. The oxidation process is more active in the presence of moisture and results in hydrated oxides (e.g.) minerals containing Fe and Mg:

$$4FeO \text{ (Ferrous oxide)} + O_2 \longrightarrow 2Fe_2O_3 \text{ (Ferric oxide)}$$

$$4Fe_3O_4 \text{ (Magnetite)} + O_2 \longrightarrow 6Fe_2O_3 \text{ (Hematite)}$$

$$2Fe_2O_3 \text{ (Hematite)} + 3H_2O \longrightarrow 2Fe_2O_3.3H_2O\text{(Limonite)}$$

The minerals which are prone to chemical attack are Feldspars, Pyroxenes and Amphiboles. The minerals that are resistant to chemical attack are Quartz and Clay minerals (e.g. Kaolinite).

Carbonation and Dissolution

Carbonation is the reaction of carbonate or carbonate ions with minerals. As rainwater falls through the atmosphere it dissolves small amounts of carbon dioxide. The atmospheric carbon dioxide when dissolved in water it forms carbonic acid.

Carbonic acid plays a significant role in chemical weathering. Carbonic acid is formed naturally in rainwater. Additional carbon di oxide is picked-up in the ground from decaying vegetation. Carbonic acid is the principal weak acid-responsible for chemical weathering:

$$2H_2O + CO_2 \longrightarrow H_2CO_3$$

This carbonic acid attacks many rocks and minerals and brings them into solution. The carbonated water has an etching effect up on some rocks, especially lime stone. The removal of cement that holds sand particles together leads to their disintegration:

$$\underset{\text{(Calcite) slightly soluble}}{CaCO_3 + H_2CO_3} \longrightarrow \underset{\text{(Ca bi carbonate) readily soluble}}{Ca(HCO_3)_2}$$

The formation would be:

$$H_2O + CO_2 = H_2CO_3 = H + 1 \text{ (carbonic acid)} + HCO_{3\text{-}1} \text{ (bicarbonate ion)}$$

The hydrogen ion in solution (H +1) is very reactive. For example, it can attack Feldspar and cause the reaction of feldspar with water (hydrolysis). Acidic rainwater is also very effective at breaking down calcium carbonate.

Calcium carbonate + carbonic acid=calcium ions + bicarbonate ions.

$$CaCO_3 + H_2CO_3 = Ca^{++} + 2HCO_{3\text{-}1}$$

Biological Weathering Processes

Unlike physical and chemical weathering, the biological or living agents are responsible for both decomposition and disintegration of rocks and minerals. The biological life is mainly controlled largely by the prevailing environment. Biological weathering comprises a group of processes that are caused by, or assisted by, the presence of vegetation, or to lesser extent animals, including root wedging and the production of organic acids.

Man and Animals

The action of man in disintegration of rocks is well known as he cuts rocks to build dams, channels and construct roads and buildings. All these activities result in increasing the surface area of the rocks for attack of chemical agents and accelerate the process of decomposition. A large number of animals, birds, insects and worms, by their activities they make holes in them and thus aid for weathering. In tropical and sub-tropical regions, ants and termites build galleries and passages and carry materials from lower to upper surface and excrete acids. Rabbits, by burrowing in to the ground, destroy soft rocks. Moles, ants and bodies of the dead animals, provide substances which react with minerals and aid in decaying process. The earthworms pass the soil, through the alimentary canal and thus bring about physical and chemical changes in soil material.

Higher Plants and Roots

The roots of trees and other plants penetrate into the joints and crevices of the rocks. As they grew, they exert a great disruptive force and the hard rock may break apart. Example: papal tree growing on walls/rocks. Some roots penetrate deep into the soil and may open some sort of drainage channel. The roots running in crevices in lime stone and marble produces acids. These acids have a solvent action on carbonates. The dead roots and plant residues decompose and produce carbon dioxide which is of great importance in weathering.

Role of Micro-organisms

In early stages of mineral decomposition and soil formation, the lower forms of plants and animals like, mosses, bacteria and fungi and actinomycetes play an important role. They extract nutrients from the rock and N from air and live with a small quantity of water. In due course of time, the soil develops under the cluster of these micro-organisms. This organism closely associated with the decay of plant and animal remains and thus liberates nutrients for the use of next generation plants and also produces CO_2 and organic compounds which aid in mineral decomposition. Organisms can assist in breaking down the rocks into sediments or soils. Lichens, fungi and other micro-organisms are the typical examples.

Mass-wasting Processes

The term mass-wasting is used for a broad category of processes whereby the earth's surface materials are pulled down on a slope by the force of gravity. Mass wasting is a geomorphic process by which soil, regolith, and rock move down a slope under the force of gravity. It is also called as mass-movement. It happens on a standing rock body or regolith. It is closely related to weathering, which is the breakdown of minerals or rocks at or near Earth's surface through physical, chemical, or biological processes. It is also related to erosion because the transport of material is done through a variety of agents like wind or flowing water.

Forces for Movements

There are two forces involved in a mass standing on a slope. One is the driving force and the other one is the resisting force. The standing mass has the driving force due to its weight. The slope and substratum have the resisting force. When the driving force exceeds the resisting force, then the mass becomes unstable and moves down or falls. When the gravitational force acting on a slope exceeds its resisting force, failure of slope occurs in holding the mass.

Slope Materials

The slope material's strength and cohesion and the amount of internal friction between material help in maintaining the slope's stability. These are collectively known as the slope's shear strength. The steepest angle that cohesion less slope can maintain the mass without losing its stability is known as its angle of repose. When a slope possesses this angle, its shear strength perfectly counterbalances the force of gravity acting upon it and the mass will not leave. When the angle of repose is more, with a little trigger or disturbance, the mass moves down rapidly.

Slow Rate

Mass wasting may occur at a very slow rate, particularly in areas that are very dry or those areas that receive sufficient rainfall such that vegetation has stabilised the surface. It may also occur at very high speed, such as in rock slides or landslides, with disastrous impacts. Varieties of mass wasting happen based on the type of mass involved, type of movement involved and the type of effect they create on a region. These movements are classified according to the speed and force of the process, from extremely slow creep to very rapid, dramatic slide or fall.

Factors Responsible

The factors that can change the potential of mass wasting include:

- Change in the slope angle.
- Weakening of the material by weathering.
- Increased water content.
- Changes in the vegetation cover.
- Overloading.

Terrestrial Slopes

Mass-wasting occurs both on terrestrial slopes and submarine slopes. Inside the seas and oceans also such mass-movements happen. Mass-movement has also been observed in Mars, Venus, and Jupiter's moon Io. Examples of rapid mass wasting include landslides and avalanches, which can be the cause of widespread death and destruction when they occur in populated areas.

Role of Water

Water plays a typical role in mass-movements. Water can increase or decrease the stability of a slope depending on the amount present not only on the overlying mass but also on the slope. Water may help to create a slippery phase on the slope. Small amounts of water can strengthen soils because the surface tension of water gives the soil a lot of cohesion. This allows the soil to resist erosion better than if it were dry. If too much water is present the water may act as a lubricant, accelerating the erosion process and resulting in different types of mass wasting like mudflows and landslides.

Sand Castle

A good example of this is to think of a sand castle, which may collapse at any time. Water must be mixed with sand in order to form the castle to keep its shape. If too much water is added the sand washes away, if enough water is not added, the sand falls and cannot keep its shape.

Types of Mass Movements

Types of mass movement are distinguished based on how the soil, regolith or rock moves down slope as a whole. The process involves two major factors as type of mass and type of movement. The type of mass involved may be bedrocks or soils. The type of movement may be a fall, slide, flow or creep.

Types of Movement Involved

Types of movement involved	Type of mass involved	
	Bedrock	Soils
Falls	Rock fall	Soil fall
Slides	Slums, slides and glides	Slums, slides and glides
Flows	Fragment flow	Earth flow, debris flow, mudflow
Complex	Complex type of movements	

Sub-categories of Mass-movements

In addition to these, sub-categories of mass-movements are classified based on the type of movement involved as:

- Planar movement.
- Rotational movement.

Types of Mass Wasting

Types of mass wasting include creep, slides, flow, topples, and falls, each with its own characteristic features, and taking place over timescales from seconds to years. The down-slope movement of material, whether it is bedrock, regolith, or a mixture of these, is commonly referred to as a landslide. Under a complex situation, all of these processes generally grade into one another, so classification of such complex processes is very difficult.

Types of Mass Movements

Ultimately the classification results into the following types of mass movements:

- Creeps.
- Slides.
- Slumps.

- Flows.
- Falls.

Creeps

Creep is a long term mass-wasting process on a downhill. In this process, the combination of small movements of soil or rock occurs in different directions over time. Creep is directed by gravity gradually downslope. The steeper the slope, the faster the creep. The creep makes trees and other shrubs bend and curve to reach the sun light. These often trigger landslides because the dirt underneath is not very strong. The trees most of the times die out because of lack of water and sun, and these rarely happen in wet climates. Caused by freezing and then thawing, or hot and then cold temperature, it causes surface soils to move up then down. Creep happens at a rate that is not noticeable to the naked eye, and it also happens in the tropical regions.

Slides

Slides are sudden downhill movement of masses of rock or sediment or soils. Slides are also called Translational Slides, as they involve only straight movement along a direction. Rock slides and debris slides result when rocks or debris slide down a pre-existing surface, such as a bedding plane, foliation surface, or joint surface. Piles of talus are common at the base of a rock slide or debris slide. Slides differ from slumps in that there is no rotation of the sliding rock mass along a curved surface.

Landslides

Landslide is a general term used to describe the downslope movement of soil, rock, and organic materials under the effects of gravity and also the landform that results from such movement. Landslides can be classified into different types on the basis of the type of movement and the type of material involved. The material in a landslide may be either a rock or soil (or both).

Slump

The term slump refers to a slipping of coherent rock material along the curved surface of a decline. Slumps are also called as Rotational Slides due to the type of movement involved. These are special category slides wherein the downward rotation of rock or regolith occurs along a concave-upward curved surface. The upper surface of each slump block remains relatively undisturbed, as do the individual blocks. Slumps leave arcuate scars or depressions on the hill slope. Slumps may be isolated or may occur in large complexes covering thousands of square meters. They often happen due to human activities. Slumps are common along the roads where slopes have been over steepened during construction. They are also common along the river banks and sea coasts, where erosion has under-cut the slopes. Heavy rains, tectonic disturbances, tremors and earthquakes can also trigger slumps.

Flows

Movement of soil and regolith resembling like a fluid behavior is called as a flow. These include avalanches, mudflows, debris flows, earth flow, lahars and sturzstroms. Water, air and ice are often involved in enabling this fluid like motion of the material.

Sediment Flows

Sediment flows occur when sufficient force is applied to rocks and regolith that they begin to flow down slope. A sediment flow is a mixture of rock, and/or regolith with some water or air. They can be broken into two types depending on the amount of water present as Slurry flows and granular flows.

Slurry Flows

Slurry flows are sediment flows that contain between about 20 and 40% water. As the water content increases above 40%, the slurry flows grade into streams. Slurry flows are considered as water-saturated flows. The Slurry Flows are further classified into:

- Solifluction.
- Debris flows.
- Mudflows.

Granular flows are sediment flows that contain between 0 and 20% water. Granular flows are possible with little or no water. Fluid-like behavior is given to these flows by mixing with air. Granular flows are not saturated with water. Each of these classes of sediment flows can be further subdivided on the basis of the velocity at which flowage occurs.

Solifluction

Solifluction is a flowage of soil mass at rates measured on the order of centimeters per year of regolith containing water. Solifluction produces distinctive lobes on the hill slopes. These occur in areas where the soil remains saturated with water for long periods of time.

Debris Flows

Debris Flows occur at higher velocities than solifluction. The velocities may range between 1 meter/hr to 100 meters/hr. It often results due to heavy rains causing saturation of the soil and regolith with water. They sometimes start with slumps and then flow downhill forming lobes with an irregular surface consisting of ridges and furrows.

Mudflows

Mudflows are very different type of mass-movements. These are a highly fluid, high velocity mixture of sediment and water that has a consistency ranging between soup-like and wet concrete. They move at velocities greater than 1 km/hr and tend to travel along valley floors. These usually result due to heavy rains in areas where there is an abundance of unconsolidated sediment that can be picked up by the streams. Thus after a heavy rain streams can turn into mudflows as they pick up more and more loose sediments on the way. Mudflows can travel for long distances over gently sloping stream beds. Because of their high velocity and long distance of travel they are potentially very dangerous types of mass-movements. Mudflows can also result from volcanic eruptions that cause melting of snow or ice on the slopes of volcanoes, or draining of crater lakes on volcanoes. Volcanic mudflows are often referred to as lahars. Some lahars can be quite hot, if they are generated as a result of eruptions of hot tephra.

Granular Flows

Granular Flows are yet another type of mass-movements. Creep is a very slow, usually continuous movement of regolith down slope. Creep occurs on almost all slopes, but the rates vary. Evidence for creep is often seen in bent trees, offsets in roads and fences, and inclined utility poles.

Earth Flows

Earth flows are usually associated with heavy rains and move at velocities between several cm/yr and 100s of m/day. They usually remain active for long periods of time. They generally tend to be narrow tongue-like features that begin at a scarp or small cliff.

Grain Flows

Grain Flows are usually formed in relatively dry material, such as a sand dune, on a steep slope. A small disturbance sends the dry unconsolidated grains moving rapidly down slope.

Debris Avalanches

Debris Avalanches are yet another type of mass-movement. These are very high velocity flows of large volume mixtures of rock and regolith that result from complete collapse of a mountainous slope. They move down slope and then can travel for considerable distances along relatively gentle slopes. They are often triggered by earthquakes and volcanic eruptions.

Topple

Topple are unique instances when huge blocks of rock pivot and fall away from a slope. Topple failures involve the forward rotation and movement of a mass of rock, earth or debris out of a slope. This kind of slope failure generally occurs around an axis (or point) at or near the base of the block of rock. A topple often results in the formation of debris or a debris cone at the base of the slope; this pile is called a talus cone. New talus cones don't have any plants growing on them. Old talus cones can have weeds and even trees on them.

Falls

Falls are major categories of mass-wasting. Falls are further categorized into two types as rock falls or debris falls. Rock falls occur when a piece of rock on a steep slope becomes dislodged and falls down the slope. Debris falls are similar, except they involve a mixture of soil, regolith, vegetation, and rocks. A rock fall may be a single rock or a mass of rocks and the falling rocks can dislodge other rocks as they collide with the cliff. A fall, including rock fall, is where regolith cascades down a slope, but is not of sufficient volume or viscosity to behave as a flow. Falls are promoted in rocks which are characterized by presence of vertical cracks or joints. Falls also occur due the undercutting action of water as well as undercutting of waves near coastal regions. They usually occur at very steep slopes such as a cliff or escarpments. The rock material may be loosened by tremors or earthquakes, rain, plant-root wedging and expanding ice.

Talus

The accumulation of rock material that has fallen down will reside at the base of the structure. It is known as talus. Because this process involves the free fall of material, falls commonly occur where there are steep cliffs. At the base of most cliffs we get the accumulation of fallen material called as talus.

Factors Influencing Mass Wasting

The major factors responsible for mass-movements are:

- Shear strength of mass.
- Slope of the land.
- Weathering and climate.
- Water content of the mass.
- Vegetation.
- Overloading.
- Geology and structure.
- Thickness of the soil cover.
- Triggering mechanism.

Triggers of Mass Wasting

Mass-wasting is a geomorphic process happening due to a triggering mechanism. Soil and regolith remain on a hillslope only while the gravitational forces are unable to overcome the frictional forces keeping the material in place. They are stable and the factor is called as Slope stability. Other Factors that reduce the frictional resistance relative to the downslope forces initiate the mass –movement. They are:

- Seismic shaking.
- Increased overburden from structures.
- Increased soil moisture and infiltration of water.

- Reduction of roots holding the soil to bedrock.
- Undercutting of the slope by excavation or erosion.
- Weathering by frost heave.
- Bioturbation.

Tectonic Causes

The most common triggering mechanisms are strong vibrations from earthquakes and excessive amounts of water from a winter snow melt or a heavy rainstorm. Volcanic eruptions, explosions, and even loud claps of thunder may also be enough to trigger a landslide if the slope is sufficiently unstable. Many avalanches, which are rapid movements of snow and ice down steep mountain slopes, are triggered by the sound of a loud gunshot, a sonic boom from a supersonic airplane, or, in rare cases, even a person's shout.

Role of Gravity

Mass wasting is fully related to gravity. Creep, slides, and flows all of these process share one thing in common, namely, that they are caused by the incessant downward pull of gravity, which moves all loose slope material downwards. Gravity-driven mass wasting processes are a subset of larger set of processes that transport weathered and unweathered earth materials. The trigger for mass-movement may come due to micro-seismic activities, tectonic dislocations and shaking of the ground by neo-tectonic forces. Such incidences may be common in places where there is an active fault on a hill slope.

Active Faults

Active faults will always create the trigger. Even if the movement is very small, the effect will be more. Huge volume of mass existing on a hill slope will be moved down within a fraction of a second after trigger. It is the most dangerous situation which is not known and unpredictable by any normal methods. Continuous seismological observations using seismometers and earth stations may help in monitoring the situations.

Overweight

The second triggering mechanism is the increase in the weight by an overburden. It may be a new structure created on the slope involving a heavy weight. Sometimes cutting the slopes for laying roads and dumping the excavated mass on another slope may increase the weight. Overloading is almost always the result of human activity and typically results from dumping, filling, or piling up of material. The additional weight created by overloading increases the water pressure within the material, which in turn decreases its shear strength, thereby weakening the slope material.

Role of Water

Water has hydraulic pressure. When water enters into the soil, it fills the pore spaces of soils and weathered zones. Due to this accumulation of water molecules on the regolith and soils, the weight increases. It also starts moving down through the pores. Water can create slippery phases between the bedrock and the overburden. The amount of water in rock or soil influences slope stability. Large quantities of water from melting snow or heavy storms greatly increase the likelihood of slope failure. The additional weight that water adds to a slope can be enough to cause mass movement. The mass-movement may occur after any slow and steady rainfall. Throughout the world, the water-induced mass-movements are more in number. Rainfall is the major triggering mechanism.

Role of Vegetation

Vegetation affects slope stability in several ways. By absorbing the water from a rainstorm, vegetation decreases water saturation of a slope's material and the resultant loss of shear strength that frequently leads to mass wasting. The removal of vegetation by either natural or human activity is a major cause of many mass movements. Removal of trees may affect the binding mechanism between overburden and the bedrock. Roots of trees have the ability to keep the mass in-tact on a slope.

Growth of Trees

Growth of trees is a slow and steady natural process and hence it never triggers a mass-movement, but does the opposite. It binds the mass with the bedrock. Due to incidences like forest fires, such unexpected conditions arise. Erosion is a major triggering mechanism. It forcefully removes the mass down a slope. Sometimes, the under-cutting action of flowing water along a slope, may move the surrounding rock bodies and regolith downwards. Similarly, cutting of slopes for highways or constructions may also lead to this situation. Weathering is another process enacting mass-movements. Weathering has the ability to disintegrate the overburden and decompose the mass both physically and chemically. These physical disintegration processes may loosen the mass. Once the compactness is disturbed, then the other factors like water saturation takes a lead role to move the mass. Though weathering is a slow process, it induces the mass to move even on a gentle slope.

Root Wedging

Root wedging is a different process. Through cracks and fractures, the roots penetrate and destabilize the rock block below. Such actions may encourage rock falls and debris falls.

Bioturbation

Bioturbation is yet another factor for mass-wasting. The role played by microorganisms may loosen the compactness of the mass on a slope.

Slope Failure

When the gravitational force acting on a slope exceeds its resisting force, slope failure (mass wasting) occurs. The resisting forces helping to maintain slope stability include the slope material's strength and cohesion, the amount of internal friction between grains, and any external support of the slope. These factors collectively define a slope's shear strength.

Slope Angle

Opposing a slope's shear strength is the force of gravity. Gravity operates vertically but has a component acting parallel to slope, thereby causing instability. All slopes are in a state of dynamic equilibrium, which means that they are constantly adjusting to new conditions. The greater a slope's angle, the greater the chance for mass wasting. A number of processes can over steepen a slope. One of the most common is undercutting by stream or wave action. This removes the slope's base, increases the slope angle, and thereby increases the gravitational force acting parallel to the slope. Excavations for road cuts and hillside building sites are another major cause of slope failure. Grading the slope too steeply, or cutting into its side, increases the stress in rock or soil until it is no longer strong enough to remain at the steeper angle and mass wasting ensues.

Weathering and Climate

Mass wasting is more likely to occur in loose or poorly consolidated slope material than in solid bedrock. As soon as solid rock is exposed at the Earth's surface, weathering begins to disintegrate and decompose it, thereby reducing its shear strength and increasing its susceptibility to mass wasting. The deeper the weathering zone extends, the greater the likelihood of some type of mass movement.

Geology and Slope Stability

Geology and Slope Stability are to be considered. The relationship between topography and the geology of an area is important in determining slope stability. If the rocks underlying a slope dip in the same direction as the slope, mass wasting is more likely to occur than if the rocks are horizontal or dip in the opposite direction.

Classification of Mass Wasting

It's important to classify slope failures so that we can understand what causes them and learn how to mitigate their effects. The three criteria used to describe slope failures are:

- The type of material that failed (typically either bedrock or unconsolidated sediment).
- The mechanism of the failure (how the material moved).
- The rate at which it moved.

The type of motion is the most important characteristic of a slope failure, and there are three different types of motion:

- If the material drops through the air, vertically or nearly vertically, it's known as a fall.
- If the material moves as a mass along a sloping surface (without internal motion within the mass), it's a slide.
- If the material has internal motion, like a fluid, it's a flow.

Unfortunately it's not normally that simple. Many slope failures involve two of these types of motion, some involve all three, and in many cases, it's not easy to tell how the material moved.

Table: Classification of slope failures based on type of material and type of motion.

Failure Type	Type of Material	Type of Motion	Rate of Motion
Rock fall	Rock fragments	Vertical or near-vertical fall (plus bouncing in many cases).	Very fast (>10s m/s)
Rock slide	A large rock body	Motion as a unit along a planar surface (translational sliding).	Typically very slow (mm/y to cm/y), but some can be faster
Rock avalanche	A large rock body that slides and then breaks into small fragments	Flow (at high speeds, the mass of rock fragments is suspended on a cushion of air).	Very fast (>10s m/s)
Creep or solifluction	Soil or other overburden; in some cases, mixed with ice	Flow (although sliding motion may also occur).	Very slow (mm/y to cm/y)
Slump	Thick deposits (m to 10s of m) of unconsolidated sediment	Motion as a unit along a curved surface (rotational sliding).	Slow (cm/y to m/y)
Mudflow	Loose sediment with a significant component of silt and clay	Flow (a mixture of sediment and water moves down a channel).	Moderate to fast (cm/s to m/s)
Debris flow	Sand, gravel, and larger fragments	Flow (similar to a mudflow, but typically faster).	Fast (m/s)

Rock Fall

Rock fragments can break off relatively easily from steep bedrock slopes, most commonly due to frost-wedging in areas where there are many freeze-thaw cycles per year. If you've ever hiked along a steep mountain trail on a cool morning, you might have heard the occasional fall of rock fragments onto a talus slope. This happens because the water between cracks freezes and expands overnight, and then when that same water thaws in the morning sun, the fragments that had been pushed beyond their limit by the ice fall to the slope below:

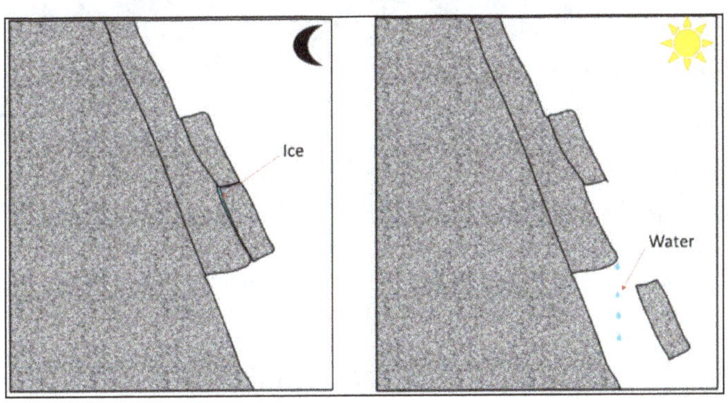

The contribution of freeze-thaw to rock fall.

A typical talus slope, near Keremeos in southern B.C., is shown in below. In December 2014, a large block of rock split away from a cliff in this same area. It broke into smaller pieces that tumbled down the slope and crashed into the road, smashing the concrete barriers and gouging out large parts of the pavement. Luckily no one was hurt.

Left: A talus slope near Keremeos, B.C., formed by rock fall from the cliffs above.
Right: The results of a rock fall onto a highway west of Keremeos.

Rock Slide

A rock slide is the sliding motion of rock along a sloping surface. In most cases, the movement is parallel to a fracture, bedding, or metamorphic foliation plane, and it can range from very slow to moderately fast. The word sackung describes the very slow motion of a block of rock (mm/y to cm/y) on a slope. A good example is the Downie Slide north of Revelstoke, B.C., which is shown in below. In this case, a massive body of rock is very slowly sliding down a steep slope along a plane of weakness that is approximately parallel to the slope. The Downie Slide, which was recognized prior to the construction of the Revelstoke Dam, was moving very slowly at the time (a few cm/year). Geological engineers were concerned that the presence of water in the reservoir could further weaken the plane of failure, leading to an acceleration of the motion. The result would have been a catastrophic failure into the reservoir that would have sent a wall of water over the dam and into the community of Revelstoke. During the construction of the dam, they tunnelled into the rock at the base of the slide and drilled hundreds of drainage holes upward into the plane of failure. This allowed water to drain out so that the pressure was reduced, which reduced the rate of movement of the sliding block. BC Hydro monitors this site continuously; the slide block is currently moving more slowly than it was prior to the construction of the dam.

The Downie Slide, a sackung, on the shore of the Revelstoke Reservoir (above the Revelstoke Dam). The head scarp is visible at the top and a side-scarp along the left side.

In the summer of 2008, a large block of rock slid rapidly from a steep slope above Highway 99 near Porteau Cove (between Horseshoe Bay and Squamish). The block slammed into the highway and adjacent railway and broke into many pieces. The highway was closed for several days, and the slope was subsequently stabilized with rock bolts and drainage holes. As shown in figure below, the rock is fractured parallel to the slope, and this almost certainly contributed to the failure. However, it is not actually known what triggered this event as the weather was dry and warm during the preceding weeks, and there was no significant earthquake in the region.

Site of rock slide at Porteau Cove: Notice the prominent fracture set parallel to the surface of the slope. The slope has been stabilized with rock bolts (top) and holes have been drilled into the rock to improve drainage (one is visible in the lower right). Risk to passing vehicles from rock fall has been reduced by hanging mesh curtains (background).

Rock Avalanche

If a rock slides and then starts moving quickly (m/s), the rock is likely to break into many small pieces, and at that point it turns into a rock avalanche, in which the large and small fragments of rock move in a fluid manner supported by a cushion of air within and beneath the moving mass. The 1965 Hope Slide was a rock avalanche, as was the famous 1903 Frank Slide in southwestern Alberta. The 2010 slide at Mt. Meager (west of Lillooet) was also a rock avalanche, and rivals the Hope Slide as the largest slope failure in Canada during historical times.

The 2010 Mt. Meager rock avalanche, showing where the slide originated (arrow, 4 km upstream). It then raced down a steep narrow valley and out into the wider valley in the foreground.

Creep or Solifluction

The very slow — mm/y to cm/y — movement of soil or other unconsolidated material on a slope is known as creep. Creep, which normally only affects the upper several centimetres of loose material, is typically a type of very slow flow, but in some cases, sliding may take place. Creep can be facilitated by freezing and thawing because, as shown in figure below, particles are lifted perpendicular to the surface by the growth of ice crystals within the soil, and then let down vertically by gravity when the ice melts. The same effect can be produced by frequent wetting and drying of the soil. In cold environments, solifluction is a more intense form of freeze-thaw-triggered creep.

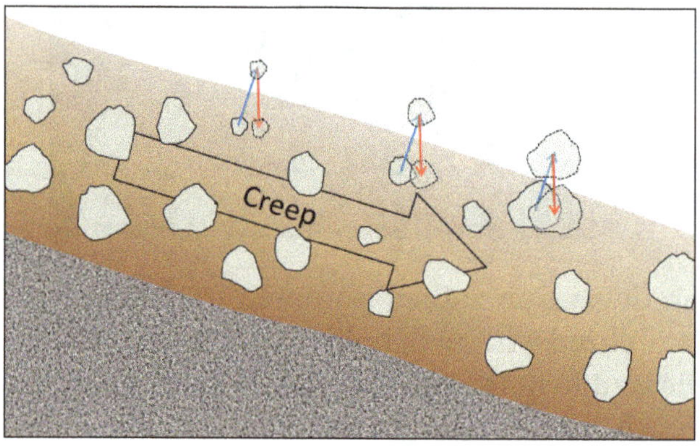

A depiction of the contribution of freeze-thaw to creep: The blue arrows represent uplift caused by freezing in the wet soil underneath, while the red arrows represent depression by gravity during thawing. The uplift is perpendicular to the slope, while the drop is vertical.

Creep is most noticeable on moderate-to-steep slopes where trees, fence posts, or grave markers are consistently leaning in a downhill direction. In the case of trees, they try to correct their lean by growing upright, and this leads to a curved lower trunk known as a "pistol butt."

Evidence of creep (tilted grave markers) at a cemetery in Nanaimo.

Slump

Slump is a type of slide (movement as a mass) that takes place within thick unconsolidated deposits (typically thicker than 10 m). Slumps involve movement along one or more curved failure surfaces, with downward motion near the top and outward motion toward the bottom. They are typically caused by an excess of water within these materials on a steep slope.

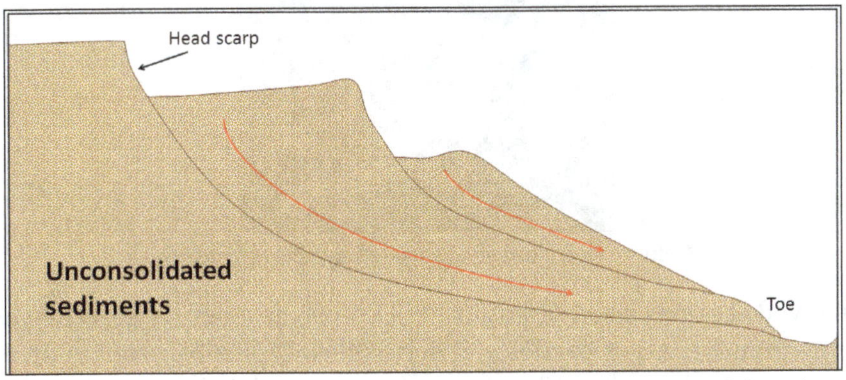

A depiction of the motion of unconsolidated sediments in an area of slumping.

An example of a slump in the Lethbridge area of Alberta is shown in figure below. This feature has likely been active for many decades, and moves a little more whenever there are heavy spring rains and significant snowmelt runoff. The toe of the slump is failing because it has been eroded by the small stream at the bottom.

A slump along the banks of a small coulee near Lethbridge, Alberta. The main head-scarp is clearly visible at the top, and a second smaller one is visible about one-quarter of the way down. The toe of the slump is being eroded by the seasonal stream that created the coulee.

Mudflows and Debris Flows

When a mass of sediment becomes completely saturated with water, the mass loses strength, to the extent that the grains are pushed apart, and it will flow, even on a gentle slope. This can happen during rapid spring snowmelt or heavy rains, and is also relatively common during volcanic eruptions because of the rapid melting of snow and ice. (A mudflow or debris flow on a volcano or during a volcanic eruption is a lahar.) If the material involved is primarily sand-sized or smaller, it is known as a mudflow, such as the one shown in below.

A slump (left) and an associated mudflow (centre).

If the material involved is gravel sized or larger, it is known as a debris flow. Because it takes more gravitational energy to move larger particles, a debris flow typically forms in an area with steeper slopes and more water than does a mudflow. In many cases, a debris flow takes place within a steep stream channel, and is triggered by the collapse of bank material into the stream. This creates a temporary dam, and then a major flow of water and debris when the dam breaks. This is the situation that led to the fatal debris flow at Johnsons Landing, B.C., in 2012. A typical west-coast debris flow is shown in figure below. This event took place in November 2006 in response to very heavy rainfall. There was enough energy to move large boulders and to knock over large trees.

The lower part of debris flow within a steep stream channel near Buttle Lake.

Fluvial Processes and Landforms

Fluvial process is the physical interaction of flowing water and the natural channels of rivers and streams. Such processes play an essential and conspicuous role in the denudation of land surfaces and the transport of rock detritus from higher to lower levels.

Over much of the world the erosion of landscape, including the reduction of mountains and the building of plains, is brought about by the flow of water. As the rain falls and collects in watercourses, the process of erosion not only degrades the land, but the products of erosion themselves become the tools with which the rivers carve the valleys in which they flow. Sediment materials eroded from one location are transported and deposited in another, only to be eroded and redeposited time and again before reaching the ocean. At successive locations, the riverine plain and the river channel itself are products of the interaction of a water channel's flow with the sediments brought down from the drainage basin above.

The velocity of a river's flow depends mainly upon the slope and the roughness of its channel. A steeper slope causes higher flow velocity, but a rougher channel decreases it. The slope of a river corresponds approximately to the fall of the country it traverses. Near the source, frequently in hilly regions, the slope is usually steep, but it gradually flattens out, with occasional irregularities, until, in traversing plains along the latter part of the river's course, it usually becomes quite mild. Accordingly, large streams usually begin as torrents with highly turbulent flow and end as gently flowing rivers.

In floodtime, rivers bring down large quantities of sediment, derived mainly from the disintegration of the surface layers of the hills and valley slopes by rain and from the erosion of the riverbed by flowing water. Glaciers, frost, and wind also contribute to the disintegration of the Earth's surface and to the supply of sediment to rivers. The power of a river current to transport materials depends to a large extent on its velocity, so that torrents with a rapid fall near the sources of rivers can carry down rocks, boulders, and large stones. These are gradually ground by attrition in their onward course into shingle, gravel, sand, and silt and are carried forward by the main river toward the sea or partially strewn over flat plains during floods. The size of the materials deposited in the bed of the river becomes smaller as the reduction of velocity diminishes the transporting power of the current.

Since the earliest days of modern applied hydraulics, engineering research has attempted to better understand sediment transportation. Because sediment particles are generally heavier than the amount of water they displace, the Archimedes principle could not be used to explain the fact that heavy sediment was capable of being lifted and transported by flowing water. Another explanation was, consequently, required. Twentieth-century research distinguishes, in this connection, between "bed load" on the one hand and "suspended load" on the other. The former is composed of the larger particles, which are either rolled or pushed along the bed of the stream or which "jump," or saltate, from the crest of one ripple to another if the velocity is sufficiently great. On the other hand, the smaller particles, the suspended sediment once picked up and lifted by the moving water, may remain in suspension for considerable periods of time and thus be transported over many kilometres.

Erosional Landforms

V Shaped Valley

Deep cutting and erosion by rapid flow of the stream carves out a valley that resembles the English letter V. The V shaped valley has a deep and narrow bottom (or valley) floor with steep valley sides. The shape is an outcome of a number of other factors at interplay; Mass movement and weathering of rocks are effective in shaping the valley.

Gorge

It is a narrow chasm with a very steep precipitous wall. These are common features found very often in mountainous regions. The Himalayas are home to a number of gorges located at different places in its ranges. The Kali Gandaki Gorge is one of the deepest gorges in the world. River Satluj enters India through a huge gorge near Shipkila pass. Brahmaputra also moves through various gorges and enters Indian Territory. Besides Himalyas the gorges have their presence across the world in most of the continents.

Gorge.

Waterfalls

It refers to a sudden descend of the flow of the stream or river caused by a variety of factors. They are very picturesque and beautiful features. Waterfalls may result due to variation in the resistance of rocks, crust deformation, changes in the sea level etc. The different mode of origin has made its classification rather difficult. However, they can broadly be classified into 5to 6 major categories taking into account their origins and distribution across the world.

Formation of Waterfalls Due to Structural Variation

Resistant Rock Lain Horizontally

When a hard and resistant rock overlies a very weak rock, the latter wears down quickly and the resistant rock shows deep undercutting. The river bed gets steepened at the point giving rise to waterfalls of larger dimension. Niagara Fall is a perfect example of this type of fall where the caprock is a strong dolomite limestone of Silurian age. The inclination of the resistant rock determines the nature and size of the waterfall. If the overlying hard rock dips downstream it results in the formation of rapids (a waterfall of small dimension). When it dips upstream it gives rise to a precipitous wall and a relatively larger waterfall.

Niagara Falls.

Vertical Falls

Granitic intrusions (or vertically arranged hard rocks) lying vertical in the course of the river are least affected by the erosion as compared to the adjoining rocks. The differential erosion accentuates the development of a steep slope causing a high water fall. Yellowstone river fall is an example of this type of fall, here a dike stands adjoining weaker rocks and Yellowstone River has made a stupendous fall when it flows over it cutting deep gorges as it descends. Such falls do not recede upstream in comparison to the falls produce on horizontal rocks because there is no undercutting of soft rocks.

Fault and Fracture Falls

Rivers flowing over faulted rock structure give rise to such type of falls. The fault scarps created during faulting cause the water to fall from a vertical height resulting in waterfalls. The Victoria Falls of river Zambezi is often cited as an example of this type of fall, because the river flows over a basaltic plateau and crosses the prominent fractures in the plateau thus giving rise to waterfall.

The Victoria Falls.

Falls due to Descend from Uplifted Highlands

When a river flows from the uplifted hard and resistant rocks (like igneous) to the plain region, It gives rise to waterfalls. Such types of falls are noticeable in the Appalachian regions where streams have developed numerous falls because of their descent into the Atlantic coastal plain. Other examples of rivers descending from plateau to the lowlands may also be included in this type of falls. E.g. the Livingstone falls, Depict the abrupt break in the flow of river Congo giving rise to an exhilarating waterfall.

Hanging Valley Falls

The main glaciated valley eroded deeper than their tributary valleys when they were formerly occupied by the glaciers. Post glaciation the glacial valleys were occupied by water and as the tributaries meet the deeply eroded and steep wall of the main glacial valleys they made waterfalls because of their sudden descent into the main valley. There are a lot of waterfalls present in the Scandinavian region. Yosemite falls in California lies in this category of waterfall. The stream (tributary of the main river) exhibits a fall over the cliff into the Merced River Valley which was previously scoured and deepened by the glacier. This deep valley lies well below the valley of its tributary thus giving rise to hanging valley fall.

Hanging Valley Falls.

Besides the classification of waterfalls on the basis of their mode of origin, there are a numerous other types of waterfalls, which are classified on the basis of their size, height and flow rate of water.

Depositional Landforms

Alluvial Fans

When a stream carrying heavy load descends from a narrow mountain valley onto a plain, it leaves behind its load in the form of a fan or a cone. This deposit at the base of a mountain occurs due to sudden drop in the velocity of the stream which cannot carry further such a heavy load on a plain region as it lacks gradient that earlier provided velocity to the stream enabling it to carry enormous load with much ease. The alluvial fans show a thick deposit at the mouth of valley and gradually descend in height away from it. The size of material deposit also show a gradual decline from large to small away from the mountain front. The slope of fans varies due to variation in the size of the grains. Where there is abundance of coarse sediments there is development of fans with steeper slopes whereas fans having finer sediments have gentler slopes. The alluvial fans in comparison to cones have a wider spread making an arc (or fan) in shape, while cones are relatively less extensive in their spread and have a steeper gradient. These two terms are often used together to denote the deposit at the mountain front. In India one can see a fan on river Kosi at the base of the Himalyas. It is a large and gently sloping fan with a width of about 140 kilimetres.

Alluvial Fan.

Floodplains

Floodplain is a very gently sloping flat region bordering the stream. It is covered with fine silt, mud sand etc. brought down by the river and deposited in the adjoining region due to its regular flooding. This regular flooding enriches the soil and makes it a very good fertile land suitable for cultivation. River flowing across a floodplain may either form meanders in its course or develop numerous channels due to deposition of the sediment load on its bed. It also gives rise to embankment known as "Natural Levees" running along it for a fairly long distance.

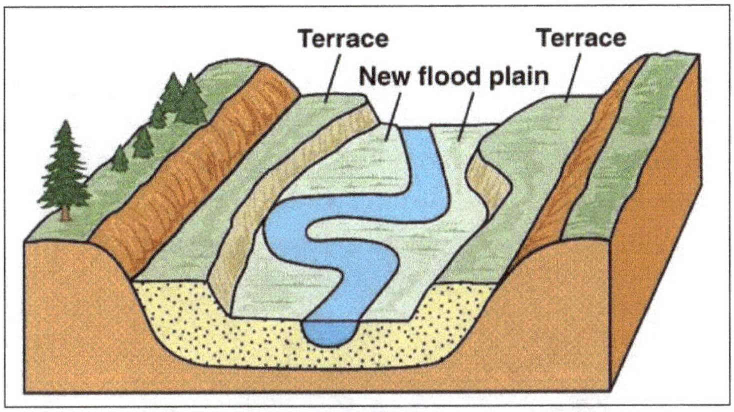

Floodplain and Levees.

Meanders

A stream flowing on a flat or a gently sloping surface seldom flows straight; it tends to take a sinuous course. While flowing it makes gentle loops in its flows commonly referred to as meander. There is no unanimity on the reason leading to formation of meanders.

The water flowing in the stream when strike the banks causes erosion at the site, which in turn leads to development of bend in the channel. The repeated striking of water and resultant erosion accentuates the bend and makes it more prominent. This prominent bend takes the form of meander. The meander represents the process of erosion and deposition going together on its banks. The outer side of the bend shows erosion and referred to as cut off bank while the inner side of the meander where the flow of the water is slow gives rise to deposition at the site of the bend known as point bar.

These two processes going on together cause the meanders to shift laterally. So meandering also leads to the formation of floodplain. As has been said the process of erosion and deposition continues it tends to accentuate the meander leading to formation of a complete circle. The river then cuts across the meander loops and takes a straighter course. The abandoned meander loops form ox bow lakes in the flood plains.

A Meanders.

Meanders and ox Bow lake.

Natural Levees

They are embankments formed on both the sides of the river. They owe their origin to the regular flooding of the river. During floods the river overflowing its bank leaves behind a good amount of material. The material due to regular occurrence consolidates into small embankments known as Natural Levees. These Levees sometimes rise very high as they grow with each successive flood.

Braided Streams

The streams when get overloaded with sediments do not carry them along and leave the excess material on the river floor in the form of sand bars. These deposits cause the stream to split into several channels. The braided stream is a common occurrence in the region which are relatively dry and arid and where the supply of water is not steady.

Braided Stream.

Alluvial Terrace

Alluvial terraces are the sites of former floodplains which are deeply eroded by the stream due to its rejuvenation. Its process of formation is explained in the following stages:

- A stream is creating its valley by erosion.
- The stream with decline in its gradient tends to deposit its load in adjoining areas creating extensive floodplain.
- Due to fall in the base level (because of fall in sea level) or upliftment of the region over which the stream flows, rejuvenation of stream takes place. The rejuvenated stream gets actively involved in downcutting of the floodplain resulting in development of terraces on both the sides. New terraces can also be created by further erosion on the newly established floodplain. Thus, lowering of Base Level either due to fall in sea level or upliftment of the land surface causes the formation of alluvial terraces.

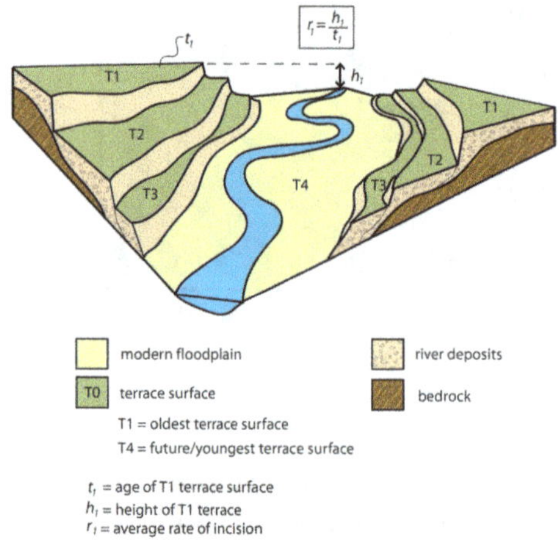

River Terrace.

Alluvial Terrace.

Deltas

Deltas are an arcuate or fan shaped feature at the mouth of the river formed by regular deposition of the sediments. The process of formation of delta begins with the deposition of sediments as soon as the river enters sea or lake. First the river would shed its coarse sediments which are heavier in comparison to the finer light particles. The finer particles are carried forward by the distributaries to some distance where they come in contact with the saline water, get coagulated and settle. It is clear that delta formation follows a series of steps. The longitudinal cross section of delta shows three depositional units:

- Bottom set beds: This is a layer of fine material on the bottom of sea or lake beyond the delta.
- Foreset beds: They are inclined beds lying between the top set beds and the bottom set beds. They are inclined because the sediments accumulate near the delta front resulting in its sliding from the frontal slope of the delta.
- Top set beds: They lie on the delta surface and as the name suggest they lie on the top of all the above mentioned beds.

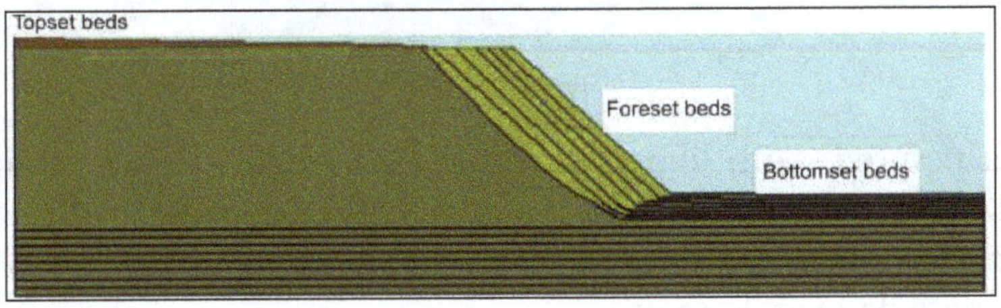

Conditions for Delta Formation

Delta formation is an outcome of a number of factors operating together with variable rate. Delta formation takes place in those parts of the coast which are relatively protected from high tides and destructive waves. So a comparatively calm sea or lake is a precondition for delta formation. The second important factor at play is the presence of abundant load in the stream which can settle at the mouth of the river. The rate of deposition of bedload should exceed the rate of its destruction. When these two conditions appear favourable delta formation takes place. Delta is not a stable feature and is regularly modified by the waves which often redistribute the deposited material at different places along the coast. Further the fluctuations in the volume of water and the amount of load a river carries also cause modification in the shape and extent of delta. Therefore, there is no uniformity in the shape and size of delta across the world, though arcuate delta is very commonly observed. The shape of the deltas can be classified into; (a) River dominated (b) Wave dominated and (c) Tide dominated. Given below are few deltas which have been categorized on the basis of their shape.

Arcuate Delta

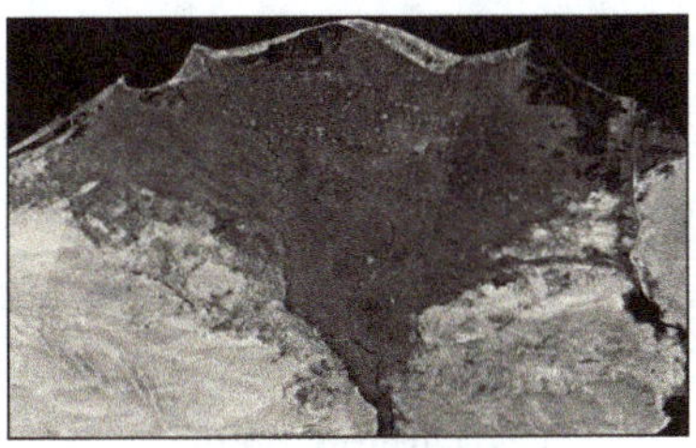

Nile Delta Source.

These Deltas resemble arc of a circle. This is the most common shape of the delta found across the world. The distributaries of the main river spread the load in large proportion as they arrive close to the sea and the deposited material coalesce to give a triangle like appearance. The Nile Delta is an ideal example of arcuate delta. It is also an example of wave dominated delta. It has barriers that enclose several lagoons. Niger delta, Ganga delta, Rhone delta, Indus delta, Po delta are all examples of arcuate delta.

Cuspate Delta

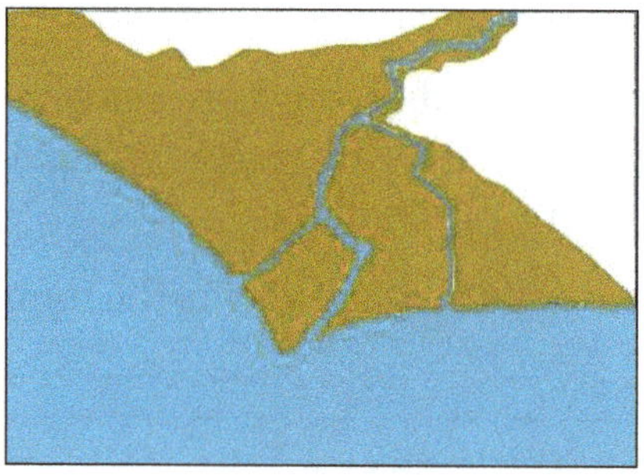

Cuspate Delta.

Cuspate delta gives a pointed tooth like appearance. It is formed on a straight shoreline where waves are in a predominant position modifying regularly the deposited sediments by redistributing it along the shoreline. The site of the river meeting the ocean looks like a protrusion in the water. Tiber River forms cuspate delta at its mouth.

Birdfoot Delta

The delta derives its name from the shape that resembles the claws of bird. The delta has its tributary channels projecting into the sea. The fine load carried by the distributaries is deposited along them in a linear form making small levees. These channels appear penetrating into the sea in narrow finger like shape. The Mississippi river is a perfect example of river dominated delta. It is a river of North America that forms grand bird foot delta when it falls in the Gulf of Mexico. This type of delta is usually formed when a large river joins a relatively shallow gulf or lake. Truncated delta, Eastuarine delta, Abandoned delta are examples of some other types of delta besides these three major types of deltas.

Mississippi delta.

Aeolian Processes

Aeolian processes, involving erosion, transportation, and deposition of sediment by the wind, occur in a variety of environments, including beaches, semi-arid and arid regions (e.g., cold and hot deserts), agricultural fields, and some terrestrial planets, notably Mars and possibly Venus. Common features of these environments are a sparse or non-existent vegetation cover, a supply of fine sediment (clay, silt, and sand size), and strong winds. Aeolian processes are responsible for the emission and/or mobilization of dust and formation of areas of sand dunes. They largely depend on other geological agents (e.g., rivers, waves) to supply sediment for transport. Despite assertions by many early workers, erosion of bedrock by the wind is not a significant process in most regions.

Aeolian processes are widespread in the Earth system, especially in low- and mid-latitude dryland environments, and also affect other environments through transport and deposition of dust. Because water is an important limiting factor in most dryland environments, these processes are strongly affected by climate changes that affect the amount, type, and temporal and spatial distribution of precipitation, which affect the supply, availability, and mobility of sediment. Dust emissions, sand transport, and dune mobility are all strongly influenced by antecedent precipitation, which affects soil moisture and vegetation cover, although specific responses are generally complex and nonlinear. Models for the response of aeolian processes to climate change on annual to decadal scales can be developed on the basis of studies of modern short-term changes in climate (e.g., drought cycles), validated by reference to late Holocene and historical climate change.

Coastal Processes and Landforms

The subject of geomorphology deals with landforms created by various geological agents. Among the geological agents, waves and currents are very important dynamic, never ending mechanisms, acting on the coastal areas. Seas and oceans are constantly sculpturing the edges of the continents and create various landforms all along the coasts. The coastal zones are very sensitive zones. The Continental shelves and near-shore areas comprise about 5% of the area of the world. In spite of this, two-thirds of the world's population thrives near the coasts, all over the world. Throughout the human history, people have been attracted to settle along the coasts. It is mainly because of the moderate climate, availability of fresh marine food, commercial and recreational opportunities and occurrence of many other natural resources.

Coastal Zones

Coasts are an interface between the lithosphere and hydrosphere. Coastal ecosystems are good habitats for innumerable flora and fauna. Time immemorial, beaches are considered to be the best places for tourism, fishing, mariculture, water-transport, recreation, sporting, social, political and community interactions. Sun bathing and souvenir collection are the regular activities of tourists along the beaches. Coastal zones are very unique and typical land masses, bordered by the seas and oceans. Coastal zones are not static but dynamic environments. They are affected by the action of waves and tides. All shore lines are subjected to both continental and oceanic processes.

Waves and Tides

Waves, tides and currents are very powerful geomorphic agents. The erosional and depositional work of the sea waves can create many spectacular landscapes along the borders of the continents. Studying the coastal landforms are interesting aspects in geomorphology. Coastal Geomorphology is a major branch of geomorphology.

Geomorphology of Coasts

Every continent or island is bordered by a long or short coastline. Coastline is the line separating the land and sea. Coastal zones are the transition zones between terrestrial and marine habitat. They form an interface between land and oceanic natural processes. Coastal areas also are varied in their topography, climate and vegetation. Some are sandy beaches, rocky shores, with or without tidal inlets. The climate of a coast are controlled by the land and sea breezes and the humidity controlled by marine water.

Waves are powerful tools for constructive and destructive activities. Hence, the geomorphology of beach, materials and processes are always not constant due to the impact of everlasting action of tides, waves and currents.

If we look at the division of a sea coast and a sea shore, the shoreline represents the actual landward limit of the seawater. This is the boundary between land and water. This limit varies with reference to location and time. Along the coasts we could notice a high tide zone and a low tide zones. These are two observable waterlines up to which the tides normally swash the coast. It happens every day. The average water level between the high tide and the low tide is considered as the mean sea level.

Divisions of Coastal Zones

Coastal belts are divided into three divisions as backshore, foreshore and offshore. The Backshore represents the beach zone starting from the limit of frequent storm wave, above high tide shoreline. This zone includes a wave cut terrace and a storm scarp. The nearshore (sometimes called the breaker zone) is where the waves break; the offshore zone is further out to sea and is beyond the influence of the waves.

Foreshore

The Foreshore region is the region between high tide water zone and low tide water zone. It includes a beach face and a beach terrace. The surf zone exists above beach terrace. At the end of the surf zone, the breaker zone starts. The foreshore may be a sandy foreshore, shingle foreshore, muddy foreshore or a rocky foreshore.

Five Major Zones

There are five major zones identified along the coastal belts as:

- Zone of wind action.
- Limit of wave actio.

- Zone of swash and backwash.
- Zone of breakers.
- Zone of shoaling waves.

Zone of Wind and Wave Action

Zone of wind action lies on the landward side and the zone of shoaling waves exist inside the sea. The limit of wave action ends along the regions of berms, which are sediment deposited zones well above the zones of swash and backwash. The greatest amount of sediment transport as beach deposit occurs within the shore and nearshore zones. Beyond the nearshore lies the offshore region.

Features of Shoreline

The typical features of the shoreline are:

- Breaker zone.
- Surf zone.
- Swash zone.
- Berms.
- Beach Face.

Breaker zone

The Breaker zone is the area where the incoming waves become unstable, raising to a peak and breaking down. Breaker zone is an important zone within which waves approaching the coastline commence breaking. The breaker zone is also part of the surf zone. The Surf zone is an important zone where the waves of translation occur after the waves break. Sand Bars are created, inside the waters, along the zone of wave breakers. The moving water masses shape the excess quantities of detritus sediments into sorted and layered deposits.

Surf Zone

This is the zone where the waves of translation occur after the waves break. Long shore currents occur in this zone, which run parallel to the coastline.

Swash Zone

This is the area where the waves backwash the materials. It is the place where up and down movement of beach materials take place.

Berms

Berms are the flat back shore areas on beaches. This is the Sun-bathing zone with wave deposited sediments. A berm is a bench-like feature containing sands carried shoreward by the swash. Landward of the berm is a belt of dunes built by wind of loose sand swept off the berms. There is summer and winter berms, both are located within the zone of wave action. During the summer, accumulation of sand takes place and it forms the summer berms. The summer berm starts from the zone of swash and backwash. Bars are created inside the waters along the zone of wave breakers. Moving water shapes excess quantities of detritus into sorted and layered deposits.

Beach Face

A beach face is the sloping section of a beach profile below the beach berm which is normally exposed to the action of the wave swash. Beach is the basic area where much of the geological processes are happening everlastingly. A Beach is the sloping portion of the coast normally existing below the berms. This area is partly exposed by the backwash of waves (swash zone). A beach is characterised by its geometric profile and the sediments comprising in it. The parameters of a beach are:

- Its geographic setting and profile.
- The beach face slope.
- Volume of sediments lying within the beaches.
- Their environmental conditions.

Beach Cycles

A beach's ability to rebuild itself makes it a formidable bastion against the sea. Whereas headlands and cliffs erode, beaches can hold their own against the anger of the sea. A beach can store sand and grow during years of good sea winds, few rains or storms and much sunshine. As the sand pump pumps the sand from the wet beach, it causes the beach to lie steeper. During years with opposite conditions, the beach can erode and lie flatter. Beaches are classified into high, low and moderate energy beaches based on these characters. Normally, high energy conditions prevail during summer months. The wave height increases after the onset of monsoon and produce significant changes in the beach morphology.

Features of a Coastline

The other major features of a coastline are:

- Sand dunes.
- Lagoons or tidal inlets.
- Estuaries.

Sand Dunes

Sand dunes of a beach, up to 3 mt elevation, are considered to be the zone of wind action. Dunes are accumulations of wind-blown sand. Although some dunes are bare, most of the dunes near a coast are vegetated with coastal plants. Such plants existing in a coastal dune help to stabilize the dune. Dunes are very dynamic geomorphic features. They are subjected to erosion during periods of high waves and accrete during normal wave conditions. During a storm or a large swell, waves attack and erode the dunes. This process, known as scarping, releases sand that was stored in the dune into the active beach. The influx of sand is often carried offshore to build sand bars, which help to attenuate the incoming wave energy.

Lagoons

Lagoons are shallow bodies of brackish or salt (sea) water partially separated from the neighboring sea by barriers of sand or shingle. The sea water can flow only through narrow openings left by the barriers. They become the coastal ponds or lakes, if they are completely detached from the sea.

Estuaries

Marine ecosystems also include the salt marshes and wetlands located along the shores and river mouths. Within the coastal zone, several unique habitats like estuaries, tidal inlets, and foreshore ecosystems are also included. At the mouth of every old stream and a river, there will an aquatic condition which includes the habitat of both fresh and salt waters. These are called as estuarine environments. Estuaries are the coastal areas where the saline waters of the ocean meet with the fresh water of streams and rivers. Estuarine habitats are usually very productive because of the accumulation of nutrients from a large catchment of a river. Estuaries offer good fishing grounds as the water column is shallow and approachable. These zones are the breeding habitat for a variety of Shrimp and Prawn species, oysters and fish. They also provide sheltered harbours for ships and their traffic. Estuarine waters are used for cooling of water in power generation plants. Various kinds of estuaries exist along the coasts of every continent in the world.

Forces Acting on Shorelines

There are the three forces which act along the shorelines. They are the sea waves, routine todes, and the longshore currents. Waves and water movement are inter-related factors on the coastal areas. Waves are undulations in the surface of a water body. Most waves are created when kinetic energy is transferred to water by the frictional stress of wind blowing over it. The rise and fall of oscillatory waves in open water reflects the circular motion of water particles. Swells are smooth, rounded waves that travel outward from a storm center or continue as broad undulations of the ocean surface after the wind dies down. Water in the breaking wave is transported toward shore as a wave of translation.

Sea Waves

Sea Waves have strength to act. There are three main factors that affect the strength of a wave:

- The strength and speed of the wind.
- The duration of the wind - this is the length of time for which the wind has blown.
- The fetch - this is the distance over which the wind has blown.

The rise and fall of oscillatory waves in open water reflects the circular motion of water particles. Swells are smooth, rounded waves that travel outward from a storm center.

Waves Classified

Sea waves are classified into two types on the basis of depth of oceanic waters as:

- Oscillatory waves (Waves in deep water).
- Translatory waves (Waves in shallow water).

From geomorphological point of view, sea waves are classified into:

- Constructive waves: Constructive waves are low energy waves that tend to arrive at the coast at a rate of less than 8 waves per minute. Constructive waves are small in height. They have a strong swash and a weak backwash. This means that constructive waves tend to deposit material and build up a beach.
- Destructive waves: Destructive waves have much higher energy and tend to arrive at the coast at a rate of more than 8 per minute. They are much larger in height than constructive waves, often having been caused by strong winds and a large fetch. Destructive waves have a weak swash but a strong backwash so they erode the beach by pulling sand and shingle down the beach as water returns to the sea. This means that less beach is left to absorb wave energy.

Longshore Drift

Longshore Drift is powerful geomorphic agents. They can erode, transport and deposit coastal sediments. As waves break on a beach, they throw water up onto the beach in an area called the swash zone. Because the waves approach at an angle, the water is thrown up at an angle. However, the water runs back down the beach under the influence of gravity perpendicular to the shore. Thus, the crashing of waves causes water to move along the beach in a step-like fashion in the direction of wave movement. This creates a longshore current. Sand is transported along the beach. This process is called as longshore drift. Longshore drift erodes and deposits sand masses continuously along the beach. The sand that is removed from one point along the beach is replaced by sand eroded from up current zones.

Tides

Tides are routine coastal processes. Nearly all marine coastlines experience the rhythmic rise and fall of sea level called tides. The daily oscillation in ocean level is a product of the gravitational attraction of the Moon and Sun on Earth's oceans and it varies in degree worldwide. Tidal action is an important force behind coastal erosion and deposition as the shoreline migrates landward and seaward. A tide is the periodic rise and fall of oceans and bodies of water connecting them. Tides are caused by the attraction of the sun and moon. Tides are a function of the sun's and moon's gravitational pull on the oceans on earth and the rotation of the earth.

Tidal Currents

Tidal Currents are responsible for mechanical sorting of sediments under the water. During a high tide water moves landward as a flood current. During low tide water recedes seaward as an ebb current.

Long-shore Currents

Breaking waves pour water into the surf zone. The water then moves parallel to the shore as an longshore current, carrying sand and other sediments along the coast, changing the shape of the coast, and forming and eroding barrier islands and spits. When the longshore current becomes sufficiently strong, it turns offshore in a fast, narrow current called a rip current. Rips are strongest and most dangerous on days with high surf. Rip currents are dangerous. Swimmers caught in a rip sometimes panic as they are carried offshore. Sediments carried along the coast by longshore currents change the shape of the coast. Where storm waves erode the beach, the sediments are carried away, to be deposited elsewhere, building new land.

Tsunamis

Seismic sea Waves called Tsunami are yet another force of oceans. A tsunami originates from the deep oceans and reaches the continents in the form of massive strong waves. These are devastating water wave generated by an undersea earthquake.

Coastal Geomorphic Processes

Oceans are bodies of dynamic water masses. Vertical and horizontal movements of water continue to happen both at the surface and at depth at all times. Over a period of time, wave action in the surf zone will tend to plane off the entire zone. This process is known as marine planation. This is a slow process. There are so many other features formed along the coastline due to various hydrodynamic actions of waves on the sea side and aerodynamic actions wind on the landside. Wind is the major cause for wave generation. The size of a wave is related to the velocity of wind and duration of the wind.

Sea Waves

Sea waves are powerful geological agents, acting from the shorelines to the coastal belts. They erode transport and deposit the marine sediments based on various factors and processes. Erosion, transportation and deposition happen on both sides of the shoreline. Coastal rocks like cliffs are also subjected to wave actions. Sea cliffs are very unique features seen in some places.

Processes of Coastal Erosion

The combined effect of waves, currents and tides result in a variety of gradational processes acting in the coastal zone. Coastal erosion happens in the form of:

- Hydraulic action.
- Abrasion.
- Attrition.
- Solution.
- Water pressure.

Hydraulic action is the impact of moving water on the coastal rocks. It is caused by the direct impact of waves on the coasts. Enormous pressures can build as water and air are compressed into the rock fractures.

Mechanisms of Erosion

The most important erosional process in the coastal region is abrasion. Abrasion (or) corrosion is a kind of erosion happening with the help of tools of erosion. In water suspension coarse sands, pebbles, cobbles and boulders are used by the waves to attack the coastal rocks. The second one is attrition. Attrition is a process in which mechanical tear and wear can break any rock mass into fragments. Mutual collision effected by backwash and rip currents are powerful tools of coastal erosion. The next process is corrosion.

Corrosion (or) Solution is the chemical alteration of rocks which are soluble and due to their contact with the seawater. Solution is locally important especially where soluble rock is exposed along the shore. Due to periodic wetting and drying a wide range of chemical processes happen on the coastal rocks which lead to both physical disintegration and chemical decomposition. Alternate freeze and thaw can also make these rocks to be easily attacked by the waves. A good number of coastal features are formed by the action of these sea waves.

Multiple Episodes Along Coasts

Coastal sediments are subject to multiple episodes of erosion, transportation and deposition, though a net seaward transport takes place on a global scale. The deep ocean floor becomes the resting place for terrestrial sediment eroded from the land. Beach drifting transports sand grains along the beach as waves strike the shore at an oblique angle. Sediment is carried landward when water rushes across the beach as swash. Sediment is carried back toward the ocean as backwash. The continual up rush and backwash carries sand in a zig-zag like movement along the shore.

Erosional Landforms

Landforms of coastal regions are classified into two major groups as erosional landforms and depositional landforms. The notable erosional landforms of the coastal areas are:

- Sea cliffs.
- Sea caves.
- Sea Arches.
- Sea stacks.
- Wave-cut notches.
- Wave-built terraces.

Sea Cliff and Caves

Wave erosion undercuts the steep shorelines and creates the coastal cliffs. A sea cliff is a vertical precipice created by waves crashing directly on a steeply inclined slope. Hydraulic action, abrasion, and chemical solution all work to cut a notch at the high water level near the base of the cliff. Constant undercutting and erosion causes the cliffs to retreat landward. Sea caves form along lines of weakness in cohesive but well-jointed bedrock. Sea caves are prominent headlands where wave refraction attacks the shore.

Sea Arches and Stacks

A sea arch forms when sea caves merge from opposite sides of a headland. If the arch collapses, a pillar of rock remains behind as a sea stack. A sea arch is a natural opening eroded out of a cliff face by marine processes. Some arches appear to have developed from surge channels, which are created by wave refraction causing the focusing of wave fronts on the side of a headland. Caves produced on either side of a promontory may become joined over time to become a tunnel and, finally, an arch. Sea arches have been regarded as ephemeral forms tending to survive over periods of just few decades to several centuries. The term sea tunnel can be used to describe a hole in the cliff line where the arch itself is considerably longer than the width of the entrance.

Wave-cut Platform

Wave-cut platforms are often most obvious at low tide when they become visible as huge areas of flat rock. The platform can only be identified at low tides or when storms move the sand. After the constant grinding and battering, eroded material is transported to adjacent bays to become beaches or seaward coming to rest as a wave-built terrace.

Depositional Landforms

The notable depositional coastal landforms are:

- Beaches.
- Spits and bars.
- Tombolo.
- Barrier islands.
- Mud Flats.

Depositional Beaches

A beach is a section of the seashore where unconsolidated sediment, or grains of worn-down rock, has collected. Unconsolidated sediment is sediment in which the individual grains are clearly separated and can move freely, like grains of rice. In contrast, consolidated sediment is sediment in which the individual grains cling together, like particles of moist flour or mud. Most beaches are composed primarily of sand (grains of quartz and other hard minerals between 0.063 mm/0.025 in. and 2 mm/0.08 in. in size), although some are composed primarily of pebbles or fragments of seashells.

The unconsolidated and small sediments that make up a beach are easily moved by the action of waves and wind. Consequently, the shape of a beach is constantly changing as sand is removed from or deposited along the shoreline. A beach is exposed to wave action along the coast.

Beaches have a characteristic morphology that changes from season to season. Beaches are fragile ribbons of sand that are frequently broken by action of nature and man. Beaches are dynamic landforms altered by wind and waves in a continual process of creation and erosion. Seasonal cycles of sand deposition and loss dramatically affect the appearance of beaches from summer to winter.

Spits and Bars

A sand spit is one of the most common coastal landforms. A sand spit is a linear accumulation of sediment that is attached to land at one end. Sand carried parallel to shore by longshore drift may eventually extend across a bay or between headlands especially where water is relatively calm. Spits are typically elongated, narrow features built to several meters high above sea level by the action of wind and waves.

Spits often form when wave energy decreases as a result of wave refraction in a bay. When the wave energy is dissipated, it will cause the sediment to accumulate, due to the loss of ability to transport the sediments by water. Spits can extend across the mouth of a bay, but wave action is usually strong enough to wash sand out to sea or be deposited in the embayment. They may curve into the bay or stretch across connecting to the other side as a baymouth bar. When the bay is closed off by a bar it becomes a lagoon. The term bar refers to a long narrow sand embankment formed by wave action.

Tombolo

A tombolo is a depositional landform in which an island is attached to the mainland by a narrow piece of land such as a spit or bar. Tombolos are formed by wave refraction. Littoral drift from an island may form a tombolo, which is a sand bar connecting the island with the mainland.

Barrier Islands

Coastlines paralleled by offshore narrow strips of sand dunes, salt marshes and beaches are known as barrier islands. Between these islands and the mainland, we can see the features like shallow lagoons, bays, or marshes. A variety of barrier-related features could be seen along the shoreline. Bay barriers are continuous barrier beaches that close off the entrance to a bay. In the upper reaches of a bay, the bayhead barrier protects lagoon or marshland. Barriers that connect headlands together along the outer reaches of an embayment are called baymouth barriers.

Barrier Spits

Barrier spits are beaches that are attached at one end to their source of sediment. Simple spits consist of narrow finger of sand with a single dune ridge that elongates in the downdraft direction. Double spits can form if drift transports sand in two directions across and inlet, or if a baymouth barrier is cut by a tidal channel. Wave refraction at the end of a spit will transport sand to form a recurved spit. Complex spits form when a plentiful supply of sediment is transported by both the ocean and bay currents. Multiple lines of dunes can be formed by wind transport of sand across the spit.

Capes

Capes are barrier islands that project into the open sea to form a right angle shoreline. These are generally large features that are exposed to wave attack on each side, but one side is accreting while the other is eroding. This produces a distinctive series of truncated dune ridges.

Mud Flats

Mud flats are formed due to the action of tidal currents. They contain silt and clays. They are exposed during low tides and are covered during high tides. In some of the exposed mud flats, after a full retreat of a sea level, plants grow in these mudflats forming salt marshes.

Deltas

These are bodies of sediments deposited by the rivers when they confluence with the seas. Deltas build outward from the shoreline at river mouths. There are three kinds of deltas as:

- Wave-dominated Deltas.
- Tide-dominated Deltas.
- River-dominated Deltas.

References

- Weathering-geology, science: britannica.com, Retrieved 17, April 2020
- Weathering-processes, the-dynamic-earth, earth-sciences: epgp.inflibnet.ac.in, Retrieved 09, March 2020
- Mass-wasting-processes, the-dynamic-earth, earth-sciences: epgp.inflibnet.ac.in, Retrieved 14, June 2020
- Classification-of-mass-wasting, geology: opentextbc.ca, Retrieved 05, May 2020
- Fluvial-process, science: britannica.com, Retrieved 18, July 2020
- Fluvial-Porcesses-and-Landforms: epgp.inflibnet.ac.in, Retrieved 20, February 2020
- Eolian-process, earth-and-planetary-sciences: sciencedirect.com, Retrieved 02, January 2020
- Coastal-processes-and-landforms, the-dynamic-earth, earth-sciences: epgp.inflibnet.ac.in, Retrieved 14, August 2020

Chapter 3

Atmosphere of the Earth

Atmosphere is the layer of gases that surround the Earth it is responsible for the various natural phenomena that make it habitable. The atmosphere has evolved over the course of the Earth's existence in both content and structure. Its structure, evolution, divisions and associated processes have been covered extensively in this chapter to provide a better understanding of the subject matter.

Earth's Atmosphere

Earth's atmosphere is a mixture of gases that surrounds our home planet. Besides providing us with something to breathe, the atmosphere helps make life on Earth possible in several ways. It shields us from most of the harmful ultraviolet (UV) radiation coming from the Sun, warms the surface of our planet by about 33 °C (59 °F) via the greenhouse effect, and largely prevents extreme differences between daytime and nighttime temperatures.

The layered structure of Earth's atmosphere is visible in this sunset view from the International Space Station.

Gases in Earth's Atmosphere

Nitrogen and oxygen are by far the most common; dry air is composed of about 78% nitrogen (N_2) and about 21% oxygen (O_2). Argon, carbon dioxide (CO_2), and many other gases are also present in much lower amounts; each makes up less than 1% of the atmosphere's mixture of gases. The atmosphere also includes water vapor. The amount of water vapor present varies a lot, but on average is around 1%. There are also many small particles - solids and liquids - "floating" in the atmosphere. These particles, which scientists call "aerosols", include dust, spores and pollen, salt from sea spray, volcanic ash, smoke, and more.

Layers of Earth's Atmosphere

The atmosphere grows thinner (less dense and lower in pressure) as one moves upward from Earth's surface. It gradually gives way to the vacuum of outer space. There is no precise "top" of the atmosphere. Air becomes so thin at altitudes between 100 and 120 km (62-75 miles) up that for many purposes that range of heights can be considered the boundary between the atmosphere and space. However, there are very thin but measurable traces of atmospheric gases hundreds of kilometers/miles above Earth's surface.

There are several different regions or layers in the atmosphere. Each has characteristic temperatures, pressures, and phenomena. We live in the troposphere, the lowest layer, where most clouds are found and almost all weather occurs. Some jet aircraft fly in the next higher layer, the stratosphere, which contains the jet streams and the ozone layer. Higher still are the mesosphere, thermosphere and exosphere. Learn about the layers of Earth's Atmosphere:

- Troposphere (where we live).
- Stratosphere.
- Mesosphere.
- Thermosphere.
- Exosphere.

Evolution of the Atmosphere

Stage-1

Earliest atmosphere is expected to have consisted of gases in the solar nebula, primarily hydrogen and possibly simple hydrides (as observed in gas giants like Jupiter and Saturn), water vapour, methane and ammonia. With dissipation of solar nebula, these gases would have escaped, partly driven off by the solar wind.

Stage-2

Second atmosphere is likely to be consisting largely of nitrogen plus carbon dioxide and inert gases, which was produced by out gassing from volcanism, supplemented by gases produced during the late heavy bombardment of Earth by huge asteroids. A major part of carbon dioxide emissions would have dissolved in water and built up carbonate sediments. An influence of life (which would have originated as early as 3.5 billion years) has to be taken into account in evolution of the atmosphere, because hints of early life forms are to be found. In fact, in the late Archaean eon an oxygen-containing atmosphere began to develop, apparently from photosynthesizing cyanobacteria which have been found as stromatolite fossils from 2.7 billion years ago.

Stage-3

Third atmosphere is the atmosphere we experience now, with increasing oxygen content. In fact, the constant re-arrangement of continents by plate tectonics would have influenced the second atmosphere through the long-term evolution of the atmosphere by transferring carbon dioxide to and from large continental carbonate stores. As per available existence, the free oxygen did not exist in the atmosphere until about 2.4 billion years ago during the Great Oxygenation Event (its appearance is indicated by the end of the banded iron formations). Before this event, any oxygen produced by photosynthesis was consumed by oxidation of reduced materials, notably iron. Molecules of free oxygen did not start to accumulate in the atmosphere until the rate of production of oxygen began to exceed the availability of reducing materials. This point signifies a shift from a reducing atmosphere to an oxidizing atmosphere. O_2 showed major variations until reaching a steady state of more than 15% by the end of the Precambrian.

Composition of the Atmosphere

About 75% of the total gas and 99% of the total water vapour of the entire atmosphere constitutes the troposphere, thus representing the 80% of total mass of the atmosphere. Following Table lists the composition of constituents of atmosphere and their volume percentage of permanent and variable gases. Nitrogen, oxygen and argon account for more than 99.96% of the atmosphere by volume. The permanent gases have virtually constant volume ratios up to an altitude of about 60 km. Although the amounts of the variable gases are small, they are extremely important in determining the atmospheric temperature.

Table: The Composition of the Atmosphere.

Permanent constituents		Variable constituents	
Constituent	% by volume	Constituent	% by volume
Nitrogen (N_2)	78.084	Water vapor (H_2O)	0-0.04
Oxygen (O_2)	20.948	Ozone (O_3)	0-12 × 10-4
Argon (Ar)	0.934	Sulfur dioxide (SO_2)[b]	0.001 × 10-4
Carbon dioxide (CO_2)	0.036	Nitrogen dioxide (NO_2)[b]	0.001 × 10-4
Neon (Ne)	18.18×10^{-4}	Ammonia (NH_3)[b]	0.004 × 10-4
Helium (He)	5.24×10^{-4}	Nitric oxide (NO)[b]	0.0005 × 10-4
Krypton (Kr)	1.14×10^{-4}	Hydrogen sulfide (H_2S)[b]	0.00005 × 10-4
Xenon (Xe)	0.089×10^{-4}	Nitric acid vapor (HNO_3)	Trace
Hydrogen (H_2)	0.5×10^{-4}	Chlorofluorocarbons	Trace
Methane (CH_4)	1.7×10^{-4}	($CFCl_3$, CF_2Cl_2	
Nitrous oxide (N_2O)[b]	0.3×10^{-4}	CH_3CCl_3, CCl_4, etc.)	
Carbon monoxide (CO)[b]	0.08×10^{-4}		

[a] After the U.S. Standard Atmosphere with the modification.

[b] Concentration near the earth's surface.

Structure of the Atmosphere

Troposphere

The troposphere, where ~90% of air mass over the Earth resides, refers to the bottom ~10 km of the atmosphere. In the troposphere, atmospheric temperature descends upward with a slope of ~10 K km−1 for dry air and ~7 K km−1 for wet air. At night, air temperature at the surface may be lower than that up to ~100 m, due to the combination of long-wave radiation of Earth and the so-called greenhouse effect. In the troposphere, numerous field campaigns have been conducted to investigate air composition over developed areas, such as North America, Europe, East Asia, Australia and New Zealand, their downwind areas, such as the Atlantic Ocean and Pacific Ocean, and remote regions, such as the Arctic and Antarctic areas. While most observations have been made near the surface, significant efforts, such as the use of balloons, flights, rockets, and satellites, have also been made to observe the air composition above, especially in recent decades. In populated developing countries, such as China and India, field campaigns have also been conducted recently to survey the chemicals responsible for air pollution, such as O_3, acid rain, and particulate matter. On a global, annual average basis, the modern tropospheric air composition excluding H_2O, CO_2, CH_4, and N_2O is listed in table below, which is termed "dry air".

Table: Dry air composition.

Dry Air	Molar Mixing Ratio
N_2	7.81E-01
O_2	2.10E-01
"Noble gases"	9.32E-03
H_2	6.00E-07
Sum	1.00E+00

1E-01 denotes 1 × 10−1, and molar mixing ratios of the noble gases He, Ne, Ar, Kr, Xe, and Rn are 5E-8, 1.5E-5, 0.93E-2, 1E-6, 5E-8, and 2E-19 respectively.

It can be seen that N_2 is the most abundant chemical, followed by O_2, and in turn by noble gases and H_2. The chemical composition of the dry air, in terms of the mixing ratio, changes little in the open atmosphere of the Earth, or as defined, though the O_2 mixing ratio is perturbed by humans, animals, plants, and crops, and may be modulated by geochemical processes. There are a number of hypotheses with regard to how the chemical composition of the dry air has arrived at its current status. For example, in the very beginning, the dry air of the Earth could have been purely CO_2, similar to the current status of Mars; biogeochemical processes might have gradually fixed carbon from the air to form fossil fuels underground and leaving O_2 in the air. The process involved is the photosynthesis in plants that converts CO_2 and H_2O into O_2, while other processes are the subject of Earth system modeling. Mixing ratios of N_2, H_2, and noble gases in the dry air are speculated to result from complex biogeochemical processes. At present levels, these gases, except Rn, have no reported adverse effects on human health, and humans and animals may have adapted to their levels in the air. As an industrial resource, N_2 is routinely used to make nitrogen fertilizers and is used as a liquid agent for small surgery, and He is used to fill balloons.

Besides the dry air, H_2O is an important component of the air in the troposphere. On one hand, it is the reservoir of precipitations that provide economic drinking water and water supplies for agricultural, industrial, and recreational purposes. On the other hand, it is a natural and the most important greenhouse gas in modern air that raises the temperature of surface air by over 30 K so that the Earth's surface is habitable for humans and animals. The mixing ratio of H_2O vapor in the troposphere ranges from <0.01 percent to a few percent, depending on elevation, latitude, longitude, surface temperature and other characteristics, such as closeness to bodies of water such as ponds, rivers, lakes, estuaries, seas, and oceans. The air may contain a small amount of liquid water as rain, cloud, fog, haze, or wet aerosol; when air is cold enough, such as in nontropical areas during winter or in the upper troposphere, it may also contain an even smaller amount of solid water as snow, hail, graupel, frost, cirrus cloud, contrails, or other icy particles suspended in the air. Table below lists typical seasonal saturated water vapor mixing ratio over the northern hemisphere, which ranges from 0.1% to 4%. Over global oceans, the relative humidity near the surface is close to 100%. Over the land, the relative humidity varies from below 5% over deserts to over 90% in coastal areas. Thus, water vapor is the third or fourth most abundant gas in surface air.

Table: Typical seasonal saturated water vapor mixing ratio.

Latitude	DJF	MAM	JJA	SON
0	0.033	0.035	0.033	0.033
15	0.041	0.035	0.037	0.035
30	0.017	0.026	0.041	0.026
45	0.006	0.013	0.026	0.015
60	0.002	0.004	0.017	0.007
75	0.001	0.001	0.007	0.003

Saturated water vapor pressure (pascals) was calculated as 610.94 × exp{17.625 × T (°C)/[T (°C) + 243.04]}. DJF, December, January, February; MAM, March, April, May; JJA, June, July, August; SON, September, October, November.

In general, the H_2O mixing ratio is higher over the tropics than over polar areas, higher in summer than in winter, higher over farmlands and forests than over deserts, and higher near the surface than further away from the surface; these phenomena reflect the facts that H_2O evaporates faster at higher temperatures and H_2O vapor is transported in the troposphere following air streams termed general circulations.

CO_2, CH_4, and N_2O are the three most important greenhouse gases in the modern troposphere, as regional and global industrialization has accelerated their increasing trends, especially in recent decades. Anthropogenic activities involving combustion harness energy from fossil fuel and biomass and emit CO_2 into the atmosphere, mostly to the troposphere, except for aviation. Globally, anthropogenic emission of CO_2 has increased dramatically since the beginning of industrialization over a century ago, and amounted to ~40 billion tons per year recently. Freshly emitted CO_2 is partly fixed by plants over the land and in surface waters, and partly dissolved into water bodies. Atmospheric CO_2 may also transform some rocks on a geochemical time scale. The remainder stays in the atmosphere, mainly in the troposphere, and raises the mixing ratio of CO_2 there. Figure below shows the annual increase of CO_2 over world oceans in the years 1996–2007. As the lifetime of CO_2 in the troposphere is an order of magnitude longer than the mixing time of tropospheric air, CO_2 is well mixed in the troposphere except at the surface with sinks or near emission sources. In fact, research has suggested that the CO_2 mixing ratio rose from ~280 ppmv in 1750 to ~310 ppmv in 1950, according to ice-core analyses, and to ~380 ppmv in 2010 based on measurements at a ground station of ~3 km ASL at the Mauna Loa Observatory in Hawaii. If anthropogenic CO_2 emission follows the current trend, the atmospheric CO_2 mixing ratio may reach 600 ppm before 2100; the exact response of atmospheric CO_2 to fossil fuel consumption depends on complex factors under active research. The increase of the atmospheric CO_2 mixing ratio has two opposite effects on humans: on one hand, a higher CO_2 mixing ratio may increase crop yields and warm up cold regions if other conditions are fixed; on the other hand, a higher CO_2 mixing ratio may have harmful consequences, such as the loss of coastal wetlands, more frequent storms or droughts, and more stagnant air near the surface.

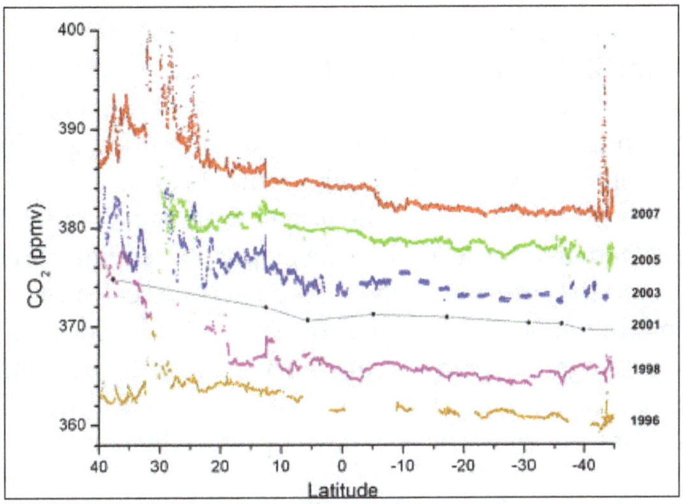

Observed atmospheric CO_2 mixing ratio.

The CH_4 mixing ratio in the troposphere is currently ~1.8 ppm, with a slightly higher mixing ratio in the northern hemisphere, where most sources are located, than in the southern hemisphere due to its relatively short lifetime (~10 years) compared with the timescale of interhemispheric air exchange (~1 year). CH_4 is the major component of natural gas, and is used widely as a clean fuel for residential, traffic, and industrial needs when available. For comparison, the CH_4 mixing ratio was estimated to be ~0.8 ppm in the middle of the eighteenth century. Tropospheric CH_4 may originate from leakages during the production, storage, transportation, and consumption of fossil fuels, and may also be emitted from rice paddies and swamps during certain periods, as well as from other sources. CH_4 is a potent greenhouse gas, e.g. with a 100-year global warming potential 21 times that of CO_2, according to the IPCC; it also contributes significantly to the photochemical production of O_3 in the troposphere on a global scale.

N_2O is rather stable in the troposphere and its current mixing ratio is ~0.32 ppm. In nature, it is a laughing gas, and is also emitted from farmlands. According to a recent survey in California, synthetic fertilizers and on-road vehicles have become dominant sources for N_2O emission there. It is estimated that tropospheric N_2O has increased by ~10% from preindustrial 1750. N_2O is a potent greenhouse gas, with a 100-year global warming potential 310 times that of CO_2, according to the IPCC.

The Troposphere as a Chemical Reactor

The troposphere extends from the surface up to the tropopause, which varies in altitude from about 8 km over the Poles to 17 km over the equator. The next higher region, the stratosphere, extends to 50 km altitude. The barometric law causes the pressure to decrease nearly exponentially with height from about 1000 hPa at the surface to 120 hPa at the tropical tropopause. The temperature decreases as well, from an average 287 K at the surface to ~200 K. The tropopause is defined by a reversal of the temperature gradient, in the stratosphere the temperature raises again.

The principal mode of heat transfer in the lower troposphere is convection, in the stratosphere it is infrared radiation. Because convection ensures rapid turbulent mixing of air, the troposphere is vertically well mixed. As vertical mixing in the stratosphere is much slower, the tropopause represents a natural boundary of transport between both regions. This makes the troposphere a fairly well-defined geochemical reservoir. It is not a well-mixed reservoir, however.

The transport of trace components generally follows the mean air currents established by the general circulation patterns in the troposphere. These are directed westward in the tropics and eastward in mid-latitudes. Transport in other directions occurs primarily via turbulent mixing processes. Turbulent transport is directed from higher to lower mixing ratios of a trace compound, that is, along the gradient of the mixing ratio. Such gradients arise because of the competition between transport and chemical removal rates. The large difference in vertical and horizontal scales, 15 versus 10,000 km in each hemisphere, makes the troposphere highly anisotropic with regard to the degree of turbulence and the rate of transport. Typical time scales within each hemisphere are 2–3 weeks for vertical mixing, 4 weeks for zonal transport, and 3 months for transport between Pole and equator. The exchange between the northern and southern hemispheres takes about 1 year. The troposphere is a reasonably well-mixed reservoir only for long-lived trace compounds, whereas others with lifetimes less than 1 yr will develop spatially nonuniform distributions.

The lifetime T of a compound is inversely proportional to its total removal rate: $T = 1/k = 1/\Sigma k_i$ [s], where the k_i are first-order rate coefficients associated with the individual processes contributing to the removal. Classical first-order processes are radioactive decay and photochemical destruction. The uptake of a trace gas by materials at the earth surface (soils and vegetation) also represents a first-order process. Most chemical reactions follow second-order kinetics. In this case, the reaction rate depends on the concentrations of two species, that of the compound being removed and that of a second reactant. First-order kinetic conditions can nevertheless be approximated by setting the concentration of the second reactant constant. This approach requires that spatial and temporal variations of its concentration are smoothed out to derive an average value. This procedure is frequently applied to estimate the lifetime of a compound in the troposphere. For longer-lived trace gases that permeate the entire troposphere, the reservoir theory holds. With the assumption that the global input rate and the total removal rate are approximately in balance, the residence time is $T = G/Q$, where G is the total trace gas content in the reservoir and $Q = \Sigma Qi$ is the global input rate resulting from the individual sources. As input and removal rates usually fluctuate considerably in space and time, it is necessary also in this model to work with suitable averages.

Stratosphere

The stratosphere contains ~9.9% of air mass over the Earth, and ranges from ~10 to ~50 km ASL with ascending temperature up to ~270 K. Due to precipitation in the troposphere, H_2O can scarcely survive through vertical transport to reach the stratosphere. In the stratosphere, O2 may be photolyzed by solar ultraviolet radiation to form ozone (O_3), which results in the so-called "O_3 layer". The O_3 layer itself absorbs solar ultraviolet radiation with a little longer wavelength, to close the Chapman cycle of O_3 formation in the natural stratosphere. As a result, solar UV radiation at the top of the stratosphere is much stronger than at the bottom of the stratosphere for wavelengths less than ~300 nm. Thus, humans and animals are effectively protected by the O_3 layer from harmful, solar UV radiation with wavelengths shorter than 300 nm.

In the modern atmosphere, chemicals such as N_2O and chlorofluorocarbons (CFCs), which decompose slowly in the troposphere, may accumulate to a significant amount and enter the stratosphere via stratosphere–troposphere exchange events. Due to strong solar radiation in the stratosphere, these chemicals photolyze to form NO and halogen radicals, which then perturb the Chapman cycle to affect the thickness of the O_3 layer. The most important observation related to the O_3 layer in the stratosphere is the so-called "O_3 hole", initially observed over Antarctica during early spring in the early 1980s. Table below lists typical seasonal column O3 over the Equator, and global distributions of column O_3; ~90% of the column O_3 stays in the stratosphere.

Table: Column O_3 in Dobson units over the Equator.

Month	O_3	Month	O_3
Jan	251	July	265
Feb	252	Aug	269
Mar	257	Sept	271
Apr	260	Oct	265
May	260	Nov	258
June	262	Dec	253

Due to strong photochemical reactions in the stratosphere, mixing ratios of surface-originated compounds, such as CH_4, N_2O, and CFCs, are lower than in the troposphere. Meanwhile, their reaction products, such as NO, OH, and halogen radicals, may be more abundant in the stratosphere than part of the troposphere.

Stratospheric Waves

The stratosphere is normally about twice as stable as the troposphere (in terms of its Brunt–Väisälä frequency, N) and permits the vertical propagation of the high-frequency waves generated in the troposphere by convection. These highest-frequency waves have phase lines that are oriented closer to the vertical than lower-frequency waves and the vertical component of their group velocity can be similar in magnitude to the horizontal component. Consequently, these waves often occur in the stratospheric air directly above convective systems. Nonetheless, as the waves propagate horizontally as well, the total horizontal area affected by high-frequency gravity waves eventually exceeds the horizontal area of the convective system that generated them.

Figure below illustrates the pattern of waves in the stratosphere above a numerically simulated isolated midlatitude convective system. (The tropopause is at approximately 12 km altitude in this case). The wave fronts are approximately circular, with the center of the fronts corresponding to the convective system itself. As is typical, the horizontal wavelength of the highest-frequency waves is smaller than the width of the convective system and corresponds to the size of individual convective updrafts. Gravity waves with horizontal wavelengths as small as 5 km have been observed directly above active convection by aircraft, consistent with those shown here. Organized MCSs comprises multiple regions of active convection and often show separate sets of circular wave fronts; each set of wave fronts is centered on a different region of active convection within the system. The result can be complex interference patterns in the wave field in the stratosphere, with additional complexity arising from source intermittency as convective regions become active and decay with time.

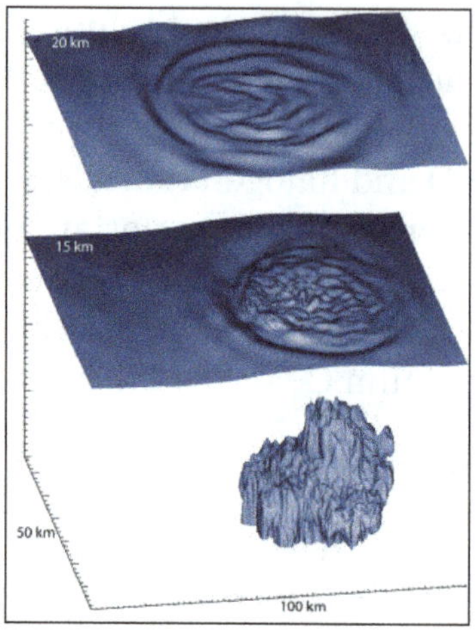

Results from a model simulation of an isolated thunderstorm. The portion of the cloud that penetrates above 10 km altitude is shown by the lower surface. Isentropic surfaces at approximately 15 and 20 km altitudes are also shown.

Convective clouds generate a spectrum of waves and lower-frequency waves that are not directly attributable to overshooting updrafts that also reach the stratosphere. For example, n = 1 and n = 2 waves are only partially reflected at the tropopause and propagate into the stratosphere as well. Typical wave periods above isolated convection can range from 10 min to a few hours, horizontal wavelengths range from ~5 to >50 km and there is a corresponding range of horizontal phase speeds. For example, figure shows an example spectrum of the vertical flux of horizontal momentum (ρuw, where ρ is the density and u and w are the perturbation horizontal and vertical velocities, respectively) at 20 km altitude above a numerically simulated midlatitude squall line. For upward propagating waves the sign of the momentum flux is equal to the sign of the phase speed relative to the mean flow, and the fluxes in this figure are consistent with upward propagating waves in an environment moving at approximately 10 m s^{-1}. In this example, peaks in the spectrum exist at phase speeds between approximately 10 and 30 m s^{-1} in magnitude relative to the 10 m s^{-1} background flow.

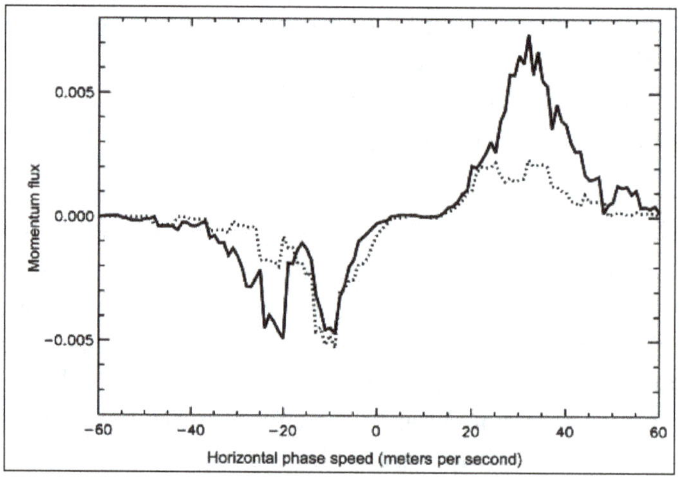

Spectrum of the vertical flux of horizontal momentum (in units of N m^{-2} per 1 m s^{-1} bin) vs horizontal phase speed above a numerically simulated squall line. Results from two simulations are shown that use horizontal grid spacing of 2 km (solid) and 125 m (dotted).

Although the examples presented earlier in figures showed symmetric wave patterns, the waves in figures show that the waves and their spectrum are different for the different propagation directions. This asymmetry occurs for two main reasons: wave generation and wave propagation.

Long-lived convective systems, like squall lines, usually feature convective updrafts that are tilted in the vertical. Studies have shown that although tilted updrafts generate gravity waves propagating in all horizontal directions, the largest amplitude waves generated are normally those that propagate in the direction of the updraft tilt. The tilted structures project onto that part of the wave spectrum more efficiently than other parts, generating an asymmetric spectrum of gravity waves.

The spectrum of waves at a given altitude in the stratosphere is also shaped by changes in wind shear and stability between that altitude and the convective wave source, i.e., wave filtering. Filtering by wind shear can create significant asymmetries in the stratospheric wave spectrum, which is closer to symmetric in the absence of shear. The shortest gravity waves are often filtered in the upper troposphere and lower stratosphere, leaving only the longer wavelength waves further aloft.

The motion of convective systems, viz. advection and propagation, complicates interpretation of the dynamics underlying the wave generation. Analyzing the wave spectrum in a frame of reference moving with the convective source is sometimes illustrative. However, choosing this reference frame is complicated by the fact that individual convective cells often propagate at a different velocity to the MCSs they are embedded within. Some studies have shown, however, that when the convective updrafts move with the mean wind at their equilibrium level, the peaks in the spectrum occur at similar phase speeds in all directions relative to that reference frame. Sheared flows with an ensemble of cloud top heights can broaden the spectrum considerably because the ensemble of wave sources moves at different speeds.

Organized convective systems, viz. systems larger than a single convective cell that persist for multiple convective overturn times, generate gravity waves with horizontal wavelengths beyond the scale of individual clouds. As convective clouds become organized, the spectrum of gravity waves broadens to encompass longer wavelengths and longer periods that are related to the spatial scale of the organized system and its lifetime. For example, regions of organized convective activity in the tropics can contain a rich spectrum of multiscale convective systems and may include individual clouds, MCSs, mesoscale convective complexes, clusters, and superclusters. Each of these systems generates gravity waves leading to a spectrum that contains notable contributions from horizontal wavelengths of 10 km through to scales beyond 1000 km. Waves with these longer wavelengths can be influenced by the Earth's rotation and are then classified as 'inertia-gravity waves.' The intrinsic period of inertia-gravity waves can be tens of hours. (The minimum intrinsic frequency of inertia-gravity waves is the local inertial frequency). Accordingly, the phase lines of inertia-gravity waves are oriented close to the horizontal and can propagate significant horizontal distances from their source. Indeed, inertia-gravity waves observed in the lower stratosphere have been traced back to convection more than 1000 km away.

Tropical cyclones and hurricanes are also notable sources of gravity waves. Patterns of vertical velocity above these systems show a combination of short-scale gravity waves from individual thunderstorms and long-scale (~100 km horizontal wavelength) waves that are associated with the evolving eye wall and spiral rainbands. Numerical simulations have found coherent spiral wave fronts above the convectively active cyclone rainbands. Notable variations in wave amplitude from these separate sources have also been linked to different stages of the cyclone intensification and decay cycle. The rotation of the wind around cyclones affects the wave propagation as well; the subsequent wave filtering, combined with different sectors of the storm featuring more active convection than others, creates considerable spatial variability in the gravity wave activity above cyclones.

Stratospheric Sulfur Compounds

Sulfate aerosols in the stratosphere are important to the albedo of the Earth. A layer of sulfate aerosols, known as the Junge layer, is found in the stratosphere at about 20 to 25 km altitude. Its origin is twofold. Large volcanic eruptions can inject SO_2 into the stratosphere, where it is oxidized to sulfate. Large eruptions have the potential to increase the abundance of stratospheric sulfate 100-fold, and the sulfate aerosols persist in the stratosphere for several years, cooling the planet. During periods without volcanic activity, the dominant source of stratospheric sulfate derives from carbonyl sulfide (COS)[8] that mixes up from the troposphere, where it originates from a variety of sources. Most sulfur gases are so reactive that they do not reach the stratosphere, but COS has a mean residence time of about 5 years in the atmosphere so about one-third of the annual production mixes to the stratosphere. Additional COS may be lofted to the stratosphere in the smoke plumes of large wildfires.

Carbonyl sulfide that reaches the stratosphere is oxidized by photolysis, forming sulfate aerosols which contribute to the Junge layer. Eventually, these aerosols are removed from the stratosphere by downward mixing of stratospheric air. The concentration of COS in the atmosphere today is higher than in preindustrial times, although it seems to have declined slightly in recent years.

Mesosphere

Based on the vertical temperature distribution in Earth's atmosphere, four semi-horizontal layers or "spheres" can be distinguished: the troposphere, stratosphere, mesosphere, and thermosphere. These layers are separated by "pauses," where no change in the temperature occurs with altitude change: the tropopause (between the troposphere and the stratosphere), the stratopause (between the stratosphere and the mesosphere), and the mesopause (between the mesosphere and the thermosphere). The stratosphere and mesosphere together are called the middle atmosphere, and their region also overlaps with the ionosphere, which is a region defined on the basis of the electric charges of the particles there.

The mesosphere, which means middle sphere, is the third layer of Earth's atmosphere, between the stratosphere, and the thermosphere. It is located from about 55 kilometers (35 miles) to 85 kilometers (54 miles) above the surface of Earth. Temperature here decreases with height, so within the mesosphere it is warmest at its lowest level (−5 °C, or 23 °F), and becomes coldest at its highest level (−80 °C, or −112 °F). Depending on latitude and season, temperatures in the upper mesosphere can be as low as −140 °C (−220 °F). The temperature in the mesosphere is lower than the temperature of the troposphere or stratosphere, which makes the mesosphere the coldest among the atmospheric layers. It is colder then Antarctica's lowest recorded temperature, and it is cold enough to freeze water vapor into ice clouds, which can be seen mostly after sunset.

Although the air in the mesosphere is relatively mixed, it is very thin, resulting in low atmospheric pressure. At this height, not only concentrations of ozone and water vapor are negligible, air in the mesosphere contains much less oxygen than in the troposphere. The mesosphere is also the layer in which many meteors burn up when they enter the earth's atmosphere, as a result of the collision with some of the gas particles present in this layer.

Thermosphere

The thermosphere is a layer of Earth's atmosphere. The thermosphere is directly above the mesosphere and below the exosphere. It extends from about 90 km (56 miles) to between 500 and 1,000 km (311 to 621 miles) above our planet.

Temperatures climb sharply in the lower thermosphere (below 200 to 300 km altitude), then level off and hold fairly steady with increasing altitude above that height. Solar activity strongly influences temperature in the thermosphere. The thermosphere is typically about 200 °C (360 °F) hotter in the daytime than at night, and roughly 500 °C (900 °F) hotter when the Sun is very active than at other times. Temperatures in the upper thermosphere can range from about 500 °C (932 °F) to 2,000 °C (3,632 °F) or higher.

The aurora (Northern Lights and Southern Lights) mostly occur in the thermosphere.

The boundary between the thermosphere and the exosphere above it is called the thermopause. At the bottom of the thermosphere is the mesopause, the boundary between the thermosphere and the mesosphere below. Although the thermosphere is considered part of Earth's atmosphere, the air density is so low in this layer that most of the thermosphere is what we normally think of as outer space. In fact, the most common definition says that space begins at an altitude of 100 km (62 miles), slightly above the mesopause at the bottom of the thermosphere. The space shuttle and the International Space Station both orbit Earth within the thermosphere.

Below the thermosphere, gases made of different types of atoms and molecules are thoroughly mixed together by turbulence in the atmosphere. Air in the lower atmosphere is mainly composed of the familiar blend of about 80% nitrogen molecules (N_2) and about 20% oxygen molecules (O_2). In the thermosphere and above, gas particles collide so infrequently that the gases become somewhat separated based on the types of chemical elements they contain. Energetic ultraviolet and X-ray photons from the Sun also break apart molecules in the thermosphere. In the upper thermosphere, atomic oxygen (O), atomic nitrogen (N), and helium (He) are the main components of air.

Much of the X-ray and UV radiation from the Sun is absorbed in the thermosphere. When the Sun is very active and emitting more high energy radiation, the thermosphere gets hotter and expands or "puffs up". Because of this, the height of the top of the thermosphere (the thermopause) varies. The thermopause is found at an altitude between 500 km and 1,000 km or higher. Since many satellites orbit within the thermosphere, changes in the density of (the very, very thin) air at orbital altitudes brought on by heating and expansion of the thermosphere generates a drag force on satellites. Engineers must take this varying drag into account when calculating orbits and satellites occasionally need to be boosted higher to offset the effects of the drag force.

High-energy solar photons also tear electrons away from gas particles in the thermosphere, creating electrically-charged ions of atoms and molecules. Earth's ionosphere, composed of several regions of such ionized particles in the atmosphere, overlaps with and shares the same space with the electrically neutral thermosphere.

Like the oceans, Earth's atmosphere has waves and tides within it. These waves and tides help move energy around within the atmosphere, including the thermosphere. Winds and the overall circulation in the thermosphere are largely driven by these tides and waves. Moving ions, dragged along by collisions with the electrically neutral gases, produce powerful electrical currents in some parts of the thermosphere.

Finally, the aurora (the Southern and Northern Lights) primarily occur in the thermosphere. Charged particles (electrons, protons, and other ions) from space collide with atoms and molecules in the thermosphere at high latitudes, exciting them into higher energy states. Those atoms and molecules shed this excess energy by emitting photons of light, which we see as colorful auroral displays.

Exosphere

The exosphere is the uppermost region of Earth's atmosphere as it gradually fades into the vacuum of space. Air in the exosphere is extremely thin - in many ways it is almost the same as the airless void of outer space.

The layer directly below the exosphere is the thermosphere; the boundary between the two is called the thermopause. The bottom of the exosphere is sometimes also referred to as the exobase. The altitude of the lower boundary of the exosphere varies. When the Sun is active around the peak of the sunspot cycle, X-rays and ultraviolet radiation from the Sun heat and "puff up" the thermosphere - raising the altitude of the thermopause to heights around 1,000 km (620 miles) above Earth's surface. When the Sun is less active during the low point of the sunspot cycle, solar radiation is less intense and the thermopause recedes to within about 500 km (310 miles) of Earth's surface.

Not all scientists agree that the exosphere is really a part of the atmosphere. Some scientists consider the thermosphere the uppermost part of Earth's atmosphere, and think that the exosphere is really just part of space. However, other scientists do consider the exosphere part of our planet's atmosphere.

Since the exosphere gradually fades into outer space, there is no clear upper boundary of this layer. One definition of the outermost limit of the exosphere places the uppermost edge of Earth's atmosphere around 190,000 km (120,000 miles), about halfway to the Moon. At this distance, radiation pressure from sunlight exerts more force on hydrogen atoms than does the pull of Earth's gravity. A faint glow of ultraviolet radiation scattered by hydrogen atoms in the uppermost atmosphere has been detected at heights of 100,000 km (62,000 miles) by satellites. This region of UV glow is called the geocorona.

Below the exosphere, molecules and atoms of atmospheric gases constantly collide with each other. However, air in the exosphere is so thin that such collisions are very rare. Gas atoms and molecules in the exosphere move along "ballistic trajectories", reminiscent of the arcing flight of a thrown ball (or shot cannonball!) as it gradually curves back towards Earth under the pull of gravity. Most gas particles in the exosphere zoom along curved paths without ever hitting another atom or molecule, eventually arcing back down into the lower atmosphere due to the pull of gravity. However, some of the faster-moving particles don't return to Earth - they fly off into space instead! A small portion of our atmosphere "leaks" away into space each year in this way.

Although the exosphere is technically part of Earth's atmosphere, in many ways it is part of outer space. Many satellites, including the International Space Station (ISS), orbit within the exosphere or below. For example, the average altitude of the ISS is about 330 km (205 miles), placing it in the thermosphere below the exosphere. Although the atmosphere is very, very thin in the thermosphere and exosphere, there is still enough air to cause a slight amount of drag force on satellites that orbit within these layers. This drag force gradually slows the spacecraft in their orbits, so that they eventually would fall out of orbit and burn up as they re-entered the atmosphere unless something is done to boost them back upwards. The ISS loses about 2 km (1.2 miles) in altitude each month to such "orbital decay", and must periodically be given an upward boost by rocket engines to keep it in orbit.

Earth's Heat Budget

The balance of incoming and outgoing heat on Earth is referred to as its heat budget. As with any budget, to maintain constant conditions the budget must be balanced so that the incoming heat equals the outgoing heat.

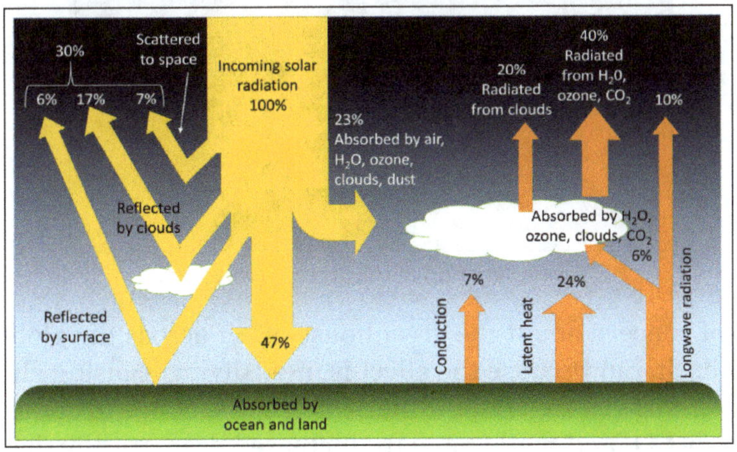

Earth's heat budget: Of all of the solar radiation reaching Earth, 30% is reflected back to space and 70% is absorbed by the Earth (47%) and atmosphere (23%). The heat absorbed by the land and oceans is exchanged with the atmosphere through conduction, radiation, and latent heat (phase change). The heat absorbed by the atmosphere is eventually radiated back into space (PW).

Of all of the solar energy reaching the Earth, about 30% is reflected back into space from the atmosphere, clouds, and surface of the Earth. Another 23% of the energy is absorbed by the water vapor, clouds, and dust in the atmosphere, where it is converted into heat. Just under half (47%) of the incoming solar radiation is absorbed by the land and ocean, and this energy heats up the Earth's surface. The energy absorbed by the Earth returns to the atmosphere through three processes; conduction, radiation, and latent heat (phase change).

Conduction is the transfer of heat through direct contact between the surface and the atmosphere. Air is a relatively poor thermal conductor (which means it is a good insulator), so conduction represents only a small part of the energy transfer between the Earth and the atmosphere; equal to about 7% of the incoming solar energy.

All bodies with a temperature above absolute zero (-273 °C) radiate heat in the form of longwave, infrared radiation. The warmed Earth is no exception, and about 16% of the original solar energy is radiated from the Earth to the atmosphere. Some of this radiated energy will dissipate into space, but a significant amount of heat will be absorbed by the atmosphere. This is the basis for the greenhouse effect. In the greenhouse effect, shortwave solar radiation passes through the atmosphere and reaches the Earth's surface where it gets absorbed. When the radiation is re-emitted by the Earth, it is now in the form of long wavelength, infrared radiation, which does not easily pass through the atmosphere. Instead, this infrared radiation is absorbed by the atmosphere, particularly by the greenhouse gases such as CO_2, methane, and water vapor. As a result, the atmosphere heats up. Without the greenhouse effect, the average temperature on Earth would be about -18 °C, which is too cold for liquid water, and therefore life as we know it could not exist.

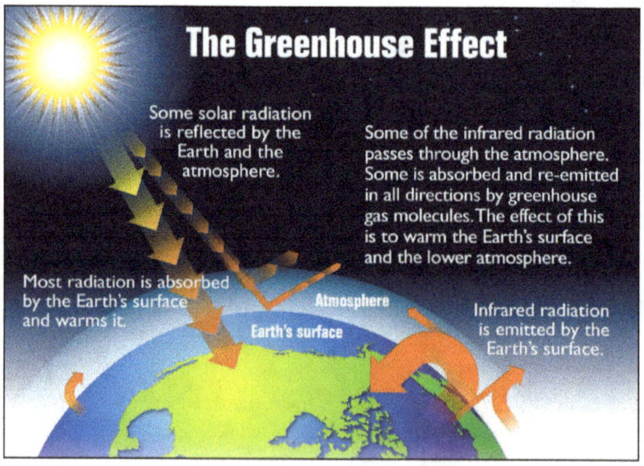

An explanation of the greenhouse effect.

There is a great deal of concern about the greenhouse effect across the globe; not because of the presence of the effect itself, but because the effect is intensifying, causing climate change or global warming. Since the Industrial Revolution the atmospheric concentrations of the major greenhouse gases, particularly CO_2 and methane, have increased dramatically due to industrialization, the burning of fossil fuels, and deforestation. At the same time, there has been rapid warming of the global climate; CO_2 concentrations have increased more than 25% and global temperature has risen by 0.50 C over the past century. Unless production of these greenhouse gases is curbed, this rapid warming trend may continue, with potentially dire consequences.

The largest pathway for heat exchange between the land or oceans and the atmosphere is latent heat transferred through phase changes; heat released or absorbed when water moves between solid, liquid, and vapor forms. Heat must be added to liquid water to make it evaporate, and when water vapor is formed, that heat is removed from the ocean and transferred to the atmosphere along with the water vapor. When water vapor condenses into rain, that heat is then returned to the oceans. The same process happens with the formation and melting of ice. Heat is absorbed by ice when it melts, and heat is released when ice forms, and these phase changes transfer heat between the oceans and the atmosphere. To complete the heat budget, the heat that is absorbed by the atmosphere either directly from solar radiation or as a result of conduction, radiation and latent heat, is eventually radiated back into space.

Differential Heating of Earth's Surface

If the Earth was a flat surface facing the sun, every part of that surface would receive the same amount of incoming solar radiation. However, because the Earth is a sphere, sunlight is not equally distributed over the Earth's surface, so different regions of Earth will be heated to different degrees. This differential heating of Earth's surface occurs for a number of reasons. First, because of the curvature of Earth, sunlight only falls perpendicularly to the surface at the center of the sphere (equatorial regions). At any other point on Earth, the angle between the surface and the incoming solar radiation is less than 90°. Because of this, the same amount of incoming solar radiation will be concentrated in a smaller area at the equator, but will be spread over a much larger area at the poles. Thus the tropics receive more intense sunlight and a greater amount of heating per unit of area than the Polar Regions.

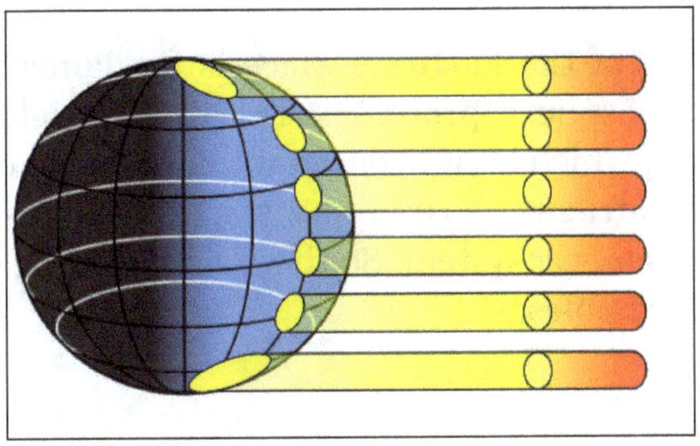

Because of the curvature of the Earth, the same amount of sunlight will be spread out over a larger area at the poles compared to the equator. The equator therefore receives more intense sunlight, and a greater amount of heat per unit of area.

The angle at which sunlight strikes the Earth contributes to differential heating of the surface in an additional way. At the poles, because of the angle at which the solar energy strikes the surface, more of the light will glance off of the surface and the atmosphere and be reflected back into space. At the equator, the direct angle with which light reaches the surface results in more of the energy being absorbed rather than reflected. Finally, the poles reflect more solar energy than other parts of the Earth because the poles have a higher albedo. The albedo refers to reflectivity of a surface. Lighter surfaces are more reflective than darker surfaces (which absorb more energy), and therefore have a higher albedo. At the poles, the ice, snow and cloud cover create a much higher albedo, and the poles reflect more and absorb less solar energy than the lower latitudes. Through all of these mechanisms, the poles absorb much less solar radiation than equatorial regions, which is why the poles are cold and the tropics are very warm.

But there is an interesting twist to this global distribution of heat. The tropical regions actually receive more radiant heat than they emit, and the poles emit more heat than they receive. We should therefore expect that the tropics will be getting continually warmer, while the poles become increasingly cold. Yet this is not the case; so what is happening? Rather than the heat remaining isolated near the equator, about 20% of the heat from the tropics is transported to the poles before it is emitted. This large scale transport of energy moderates the climates at both extremes.

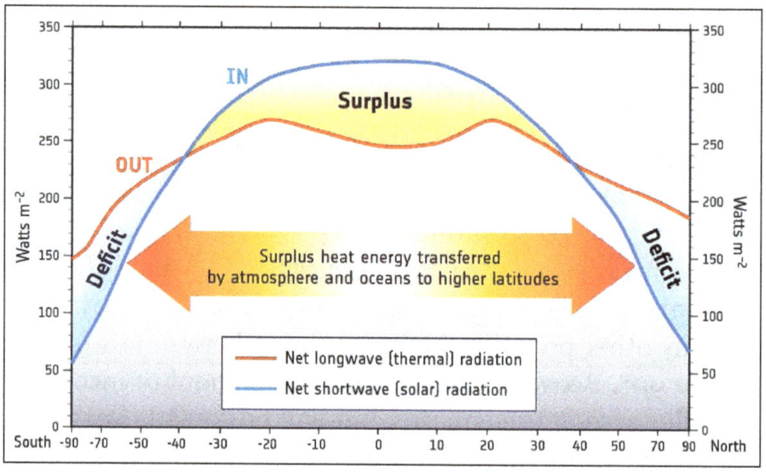

The balance between heat gain and heat loss as a function of latitude. Excess heat received near the equator is transferred towards the poles.

The idea of differential heating of the Earth's surface is fundamental to understanding a wide range of oceanographic and atmospheric processes. This differential heating leads to atmospheric convection which creates winds, which blow over the water and create waves and surface currents, and these currents influence nutrient distribution, which promotes primary production, which then supports the rest of the ocean ecosystem. So there's a lot riding on the simple fact that more light reaches the tropics than the poles.

References

- Earths-atmosphere, shortcontent: scied.ucar.edu, Retrieved 19, July 2020
- Thermal-structure-of-atmosphere, meteorology-climatology, earth-sciences: epgp.inflibnet.ac.in, Retrieved 03, March 2020
- Troposphere, earth-and-planetary-sciences: sciencedirect.com, Retrieved 07, August 2020
- Stratosphere, earth-and-planetary-sciences: sciencedirect.com, Retrieved 14, March 2020
- Mesosphere, atmospheric-and-space-sciences-atmosphere, atmosphere-and-weather, earth-and-environment: encyclopedia.com, Retrieved 09, July 2020
- Thermosphere-overview: scied.ucar.edu, Retrieved 23, January 2020
- Exosphere-overview: scied.ucar.edu, Retrieved 02, February 2020
- Earths-heat-budget, webboceanography: rwu.pressbooks.pub, Retrieved 16, June 2020

Chapter 4

Oceanography and its Types

Oceans cover more than 70 percent of the surface of the Earth. The study of the different features of the ocean is known as oceanography. It includes physical, biological, chemical, geological and coastal oceanography. This chapter closely examines the key oceanography to provide an extensive understanding of the subject.

Oceanography

Oceanography is the study of the physical, chemical, and biological features of the ocean, including the ocean's ancient history, its current condition, and its future. In a time when the ocean is threatened by climate change and pollution, coastlines are eroding, and entire species of marine life are at risk of extinction, the role of oceanographers may be more important now than it has ever been.

Indeed, one of the most critical branches of oceanography today is known as biological oceanography. It is the study of the ocean's plants and animals and their interactions with the marine environment. But oceanography is not just about study and research. It is also about using that information to help leaders make smart choices about policies that affect ocean health. Lessons learned through oceanography affect the ways humans use the sea for transportation, food, energy, water, and much more. For example, fishermen with the Northwest Atlantic Marine Alliance (NAMA) are working with oceanographers to better understand how pollutants are reducing fish populations and posing health risks to consumers of the fish. Together, NAMA and ocean scientists hope to use their research to show why tighter pollution controls are needed.

Oceanographers from around the world are exploring a range of subjects as wide as the ocean itself. For example, teams of oceanographers are investigating how melting sea ice is changing the feeding and migration patterns of whales that populate the ocean's coldest regions. National Geographic Explorer Gabrielle Corradino, a North Carolina State University 2017 Global Change Fellow, is also interested in marine ecosystems, though in a much warmer environment. Corradino is studying how the changing ocean is affecting populations of microscopic phytoplankton and the fish that feed off of them. Her field work included five weeks in the Gulf of Mexico filtering seawater to capture phytoplankton and protozoa—the tiniest, but most important, parts of the sea's food chain.

Of course, oceanography covers more than the living organisms in the sea. A branch of oceanography called geological oceanography focuses on the formation of the seafloor and how it changes over time. Geological oceanographers are starting to use special GPS technology to map the seafloor and other underwater features. This research can provide critical information, such as seismic activity, that could lead to more accurate earthquake and tsunami prediction.

In addition to biological and geological oceanography, there are two other main branches of sea science. One is physical oceanography, the study of the relationships between the seafloor, the coastline, and the atmosphere. The other is chemical oceanography, the study of the chemical composition of seawater and how it is affected by weather, human activities, and other factors.

About 70 percent of Earth's surface is covered by water. Nearly 97 percent of that water is the saltwater swirling in the world's ocean. Given the size of the ocean and the rapid advancements in technology, there is seemingly no end to what can and will be uncovered in the science of oceanography.

Diver Jeremy Stewart is conducting biological oceanography, which is the study of the ocean's plants and animals and their interactions with the marine environment, by suctioning amphipods (shrimp-like crustaceans) from the ice floe.

Oceanographic Observations

Oceanographic investigations, traditionally, are both limited and strictly dependent by time and space scales and by sensors technological developments. To observe oceanographic phenomena and processes, several and different methods can be utilized providing different results. Even if some methods still have remained almost unvaried, even if they have been improved and perfection (as for example the continuous plankton recorder, CPR), in many cases, technological developments lead us to a noteworthy advancing of the knowledge of processes and phenomena, formerly undetectable and just intuitable.

Platforms and Sensors

Relating to a defined phenomenon, observation has to be carried out with the appropriated method in order to detect processes and trends of such phenomena. To this aim, platforms and sensors have to be choosing as in respect to the time and space scale of the phenomena, as taking into account the nature (e.g. chemical, physical or biological) of what we are looking to. The environment (e.g. shallow water, deep sea or coastal areas) is also fundamental to the choice of the adequate sampling sensors and operative methods. Many physical, chemical and biological aspects can be described by means of punctual observations, e.g. standalone moored instruments or oceanography buoys and multipurpose platforms. Also, dedicated oceanographic surveys can be set up in order to study a particular aspect of a defined marine area.

In order to reduce operative costs and to enhance spatial resolution of data for water column characterization, towed vehicles can be utilized to continuous measure of physical, chemical and biological variables, along water column and along horizontal trajectories. Also the use of expendable probes and the ship of opportunity are fundamental to this aim. From an operative point of view, observation platforms can be divided basically into the following categories:

- Ships (Research Vessels RV).
- Mooring.
- Underwater vehicles and lowered devices (Remote Operated Vehicles ROV; Towed Vehicles; Crewed Deep Submersible Vehicles; Autonomous Underwater Vehicles; Gliders; Lagrangian buoys; Expendable probes; Drifters).
- Voluntary Observing Ships (VOS).
- Remote sensing platforms (Airplanes and Satellites).

Ships

Oceanographic research vessels can be considered shipborne platform as they can be utilized to conduct many kind of scientific investigations, such as mapping and charting, marine biology, fishery, geology and geophysics, physical processes, marine meteorology, chemical oceanography, marine acoustics, underwater archaeology, ocean engineering, and related fields.

The United Nations International Maritime Organization (IMO) has established a category of Special Purpose Ship which includes ships engaged in research, expeditions, and survey. Scientific personnel are also defined as "persons who are not passengers or members of the crew, who are carried on board in connection with the special purpose". The design of a modern Research Vessel can vary significantly in respect to the research activities. Typical characteristics can be as follow described:

- RV has to be as general as possible in order to allow multidisciplinary studies and researches.
- The size is determined by the requirements, but the length-over-all should not exceed 100 m LOA.
- Speed of 15 knots cruising should be sustainable through sea state 4 (1.25–2.5 m) and the sea keeping should be able to maintain science operations in the following speeds and sea states:
    - 15 knots cruising through sea state 4 (1.25– .5 m).
    - 13 knots cruising through sea state 5 (2.5–4 m).
    - 8 knots cruising through sea state 6 (4–6 m).
    - 6 knots cruising through sea state 7 (6–9 m).
- Work environment: Lab spaces and arrangements should be highly flexible to accommodate large, heavy, and portable equipment.
- Suite of modern cranes, in order to reach all working deck areas and offload vans and heavy equipment, and in order to work close to deck and water surface.

- Oceanographic winches permanently installed should provide a wire monitoring systems with inputs to laboratory panels allowing local and remote data and operational controls.

The above general characteristics can be modulating relating to the particular purpose of the vessels itself. Therefore it is more convenient and appropriate divides RV in the following categories:

- General Purpose Vessels: Multidiscipline Ships represent the classic oceanographic research vessels and are the dominant class in terms of numbers today. Current and future multidiscipline oceanographic ships are characterized by significant open deck area and laboratory space. Also accommodations for scientific personnel are greater than for single purpose vessels due to the larger science parties carried. Flexibility is an essential feature in a general purpose research vessel.
- Mapping and Charting Vessels: This group of ships was probably the earliest oceanographic vessels, traditionally involved in exploration voyage. Surveys were carried out using wire sounding, drags, and launches. Survey vessels are also characterized by less deck working space than general purpose vessels. Modern survey vessels, however, are often expected to carry out other scientific disciplines. Winches, cranes and frames can be observed on these ships.
- Fisheries Research Vessels: Fisheries research generally includes studies on environment, stock assessment, and gear testing and development. The first of these are carried out by traditional surveys to collect biological, physical, and chemical parameters of sea surface and water column as well as geological information's. These surveys can be accomplished from a general purpose oceanographic research vessel.
- Geophysical Research Vessels: The purpose of marine geophysical research vessels is to investigate the sea floor and sub-bottom, oceanic crust, margins, and lithosphere ranging from basic research of the Earth's crust to resources exploration. The highly specialized design requirements for a full-scale marine geophysics ship usually preclude work in other oceanographic disciplines.
- Polar Research Vessels: Polar research vessels are defined by their area of operations. The special requirements defining a polar research vessel include increased endurance, usually set at 90 days, helicopter support, and special provisions for cold weather work, such as enclosed winch rooms and heated decks, and icebreaking capability.
- Support Vessels: These include vessels that support submersibles, ROVs, buoys, underwater habitats, and scientific diving.
- Other Classes of Oceanographic Research Vessels: Into this category it can be include research ships which serve other purposes as ocean drilling and geotechnical ships, weather ships, underwater archaeology, and training and education vessels.

Moorings

Much of what we know about the oceans processes is the result of ship-based expeditionary science, dating back to the late 19th century. It is clear that, to answer many important questions about oceans and Earth science, it is necessary a co-ordinated research effort based on long term investigations.

One of the most important aspects of sea investigation, in fact, deals with the possibility to obtain continuous data from fixed stations or from a net of fixed stations, fundamental for forecasting systems (coupled with meteorological data and time data series).

Mooring platforms, both for upper and deep water, allow continuous observations of phenomena of very different disciplines like geophysical and biological once. Worldwide initiatives and programs, such as European Seafloor Observatory Net (ESONET NoE) and the European Multidisciplinary Seas Observatory (EMSO) have been developed in order to increase the capacity in the research, with the purpose of better understanding of physical, geological, chemical, ecological, biological and microbial processes, that take place in the oceans: from the surface down to the highest depth. Much of our latest knowledge, in fact, stems from studying the seafloor: its morphology, geophysical structure, and characteristics, and the chemical composition of rocks collected from the ocean floor.

Furthermore, in the late 1970s, at the mid-ocean ridge (MOR) crest, deep sea biological observations led to the discoveries of deep sea 'black smoker' hydrothermal vents and to their chemosynthetic-based communities. Such discovery changed the biological sciences, provided a quantitative context for understanding global ocean chemical balances. Deep sea observatory is also fundamental in the study of the physical oceanography of the global ocean water masses and their chemistry and dynamics.

Traditionally, fixed platforms consist of one surface buoyant unit moored on sea bottom by means of an instrumented chain. Most often surface unit is equipped with meteorological sensors, while along the chain (from the surface till the bottom) are mounted many different sensors in order to collect water column variables.

A mooring design can vary relating both to the kind of investigation and to physical characteristics of the environment of destination (coastal or open ocean). Fixed measurement platforms have the following features and play the following roles:

- Standalone data collecting system allowing real time or quasi-real time data acquisitions and transmission: Surface platforms are traditionally able to store sampled data into internal storage devices, and most often they are equipped with a communication system that allows data transmission to data centers. Typical systems involve satellite (open-ocean), GSM or H3G phone systems but also radio, ethernet or LAN communication can be utilized.

- Operational forecasting system: Continuous data coming from a buoy or a buoy network provide fundamental input for ocean and weather forecasting models.

- Remote sensing calibration system: In situ data can represent the sea-truth for remote sensing as they are provided from standard and calibrated instruments. Because of the high maintenance possibility and the continuous data control, fixed platforms and mooring buoys can serve successfully to this aim. The calibration of a moored buoy, for example, can be ongoing, with retrieval at certain intervals to check sensor degradation and biofouling.

- Data network: The importance of realizing a network of buoys is related to the possibility to put all the information of each platform together, in order to provide an even more rich data base. It is necessary not only for the oceanographic research and monitoring, but also for climate and global change investigations.

Following there is reported some of important buoy networks:

- An important example is represented by the United States National Data Buoy Center (NDBC). In the 1960's, there were approximately 50 individual United States programs conducted by a variety of ocean-oriented agencies. These programs were consolidated into the National Data Buoy Center under the control of the US Coast Guard. Nowadays the United States NDBC manages the development, operations, and maintenance of the national data buoy network. It serves as the NOAA focal point for data buoy and associated meteorological and environmental monitoring meteorological/environmental data in real time from automated observing technology. It provides high quality systems.

It also manages the Volunteer Observing Ship (VOS) program to acquire additional meteorological and oceanographic observations supporting the NOAA National Water Service (NWS) mission requirements. It operates the NWS test center for all surface sensor systems. It maintains the capability to support operational and research programs of NOAA and other national and international organizations.

- The Australian National Moorings Network of the integrated marine Observing System comprises a series of national reference stations and regional moorings designed to monitor particular oceanographic phenomena in Australian coastal ocean waters. Nine national reference stations are located around the Australian coasts in order to provide a baseline information's, decadal time series of the physical and biogeochemical properties of Australia's coastal seas, to inform research into ocean change, climate variability, ocean circulation and ecosystem responses.

Each National Reference Station consists of a mooring with sensors for conductivity, temperature, depth, fluorescence, dissolved oxygen, photosynthetically available radiation (PAR), fluorescence and measurement of turbidity at three depths: the surface, seabed and an intermediate depth. On the seafloor, acoustic doppler current profilers (ADCPs) are also deployed. All reference stations telemeter a reduced data set via Iridium satellite for real time monitoring. Boat-based water sampling is also undertaken at each of the reference stations on a monthly basis. These samples are analysed for nutrients, plankton species, both visibly and genetically, and pCO_2.

A number of the National Reference Station moorings are also equipped with passive acoustic listening arrays, containing sea noise loggers to record sounds in the ocean. Furthermore, three National Reference Stations are equipped with three instruments determining surface CO_2, dissolved oxygen, temperature and salinity.

Regional moorings can measure physical, chemical and biological parameter of sea water. These moorings can hold a range of instrumentation including acoustic doppler current profilers (ADCPs), Water Quality Meters (WQMs), fluorometers, instruments to measure turbidity, dissolved oxygen, photosynthetically active radiation (PAR), nutrients, pCO_2, dissolved inorganic carbon, total alkalinity, as well as imaging flow cytometetry, spectroradiometry, profiling conductivity-temperature-depth instrumentation, laser In-situ scattering and transmissometry.

The regional moorings monitor the interaction between boundary currents and shelf water masses and their consequent impact upon ocean productivity and ecosystem distribution and resilience. Operation of the moorings network facility is coordinated nationally and distributed between several sub-facilities.

- The Italian National Mareographic Net (RMN) was set up by the National Mareographic System. It consists of 33 measuring stations uniformly positioned throughout the entire national seas. The main feature of Italian RMN is the measure of sea level. Each buoy is equipped with a sea level microwave sensor coupled with another one level sensor with back-up function. In addiction there is the traditional ultrasound hydrometric sensor working since 1998. Each buoy also carries a meteorological station and a sea water temperature sensor. Ten buoys are equipped with a multiparametric probe in order to measure temperature, pH, conductivity and redox parameters.

All data collected are available for historical series updates, real time observations, astronomical tides forecast and for scientific investigations.

Other important examples are represented by the deep sea observatories:

- The Circulation Obviation Retrofit Kit (CORK) is a seafloor observatory that measures pressure, temperature, and fluid composition, important parameters for the study of the dynamics of deep-sea hydrologic systems. CORKs are installed by the International Ocean Drilling Program (IODP) for measurements over long periods of time (months to years).
- The MARS Observatory of Monterey Bay Aquarium Research Institute (MBARI) is the first deep-sea ocean observatory offshore of the continental United States and consists of a metal pyramid on the seafloor off the coast of Central California at 900m depth. Working since 2008, it is involved in the FOCE experiment (Free-ocean carbon dioxide enrichment) to study the effects of increased carbon dioxide concentrations in seawater on marine animals. The heart of observatory consists of two titanium pressure cylinders packed with computer networking and power distribution equipment. These cylinders are nested within a protective metal pyramid on the deep seafloor. This central hub is connected to shore by a 52-kilometer-long cable that can carry up to 10,000 watts of power and two gigabits per second of data. Most of the cable is buried a meter below the seafloor.
- The Ocean Observatory Initiative (OOI), a project funded by the National Science Foundation, is planned as a networked infrastructure of science-driven sensor systems to measure the physical, chemical, geological and biological variables in the ocean and seafloor. The OOI will be one fully integrated system collecting data on coastal, regional and global scales. Three major Implementing Organizations are responsible for construction and development of the overall program. Woods Hole Oceanographic Institution and its partners, Oregon State University and Scripps Institution of Oceanography are responsible for the coastal and global moorings and their autonomous vehicles. The University of Washington is responsible for cabled seafloor systems and moorings. The University of California, San Diego, is implementing the cyberinfrastructure component. Rutgers, The State University of New Jersey, with its partners University of Maine and Raytheon Mission Operations and Services, is responsible for the education and public engagement software infrastructure.

The OOI will consist of six arrays with 56 total moorings and 763 instruments. Moored platforms provide oceanographers the means to deploy sensors at fixed depths between the seafloor and the sea surface and to deploy packages that profile vertically at one location by moving up and down along the mooring line or by winching themselves up and down from their point of attachment to the mooring. An oceanographic mooring is anchored to the sea floor by a mooring line extending upward from the anchor to one or more buoyant floats, which can be located in the water column or at the sea surface.

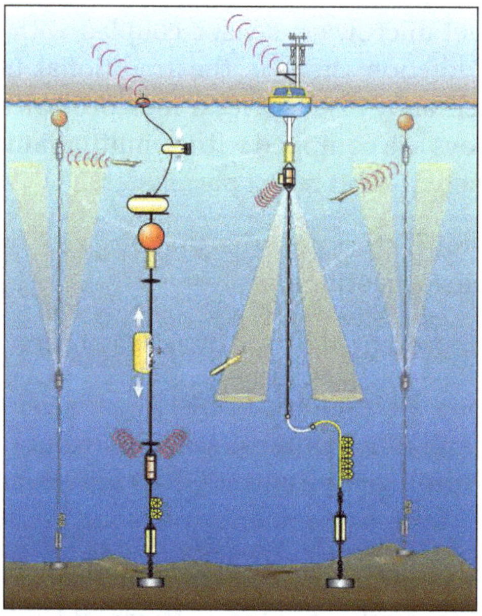

Ocean Observatory Initiative moored platforms.

Underwater Vehicles and Lowered Devices

Underwater vehicles can be considered as platforms that combine the advantages of ship-mounted instruments collecting data while moving, and a lowered device that profiles the water column. Underwater Vehicles can be divided into three categories related with their moving capabilities: towed, remote operated and autonomous. Traditionally they are called Towfish or Towed body, ROV and AUV respectively.

Towed Vehicles

A towed vehicle system has three main components: the vehicle, the tow cable and a winch. Vehicles, with their instrument payload, represent the measurement platforms. The cable, whenever double-armored (electromechanical) can be considered the principal part of the towfish. Cable is responsible for power and data transmission, but, above all, because of its total cross sectional area, cable drag dominates the performance of the system.

Towed-vehicle systems using electromechanical cables usually require a special winch with accurate spooling gear and slip-rings to make the electrical connection to the rotating drum. The faired cable has a large bending radius and can only be wound onto the winch drum in a single layer. If the towed vehicle system uses wire rope for the tow cable, then the winch can be a standard type.

Towfish payload can be different one to each other relating to the field of investigation. Typical instrument payloads can be CTDs added with specific sensor (i.e. pH and fluorometer) required for traditional hydrological studies. Acoustic devices and biological sampler can be mounted on "ad hoc" designed towed body.

One of the most important towed vehicles for biological sampling was designed by Sir Alister Hardy in 1925, and until now it is remained relatively unchanged in its sampling mechanism. Called Continuous Plankton Recorder, (CPR) this towfish is able to collect and stock zooplankton samples continuously while it is towed from a ship.

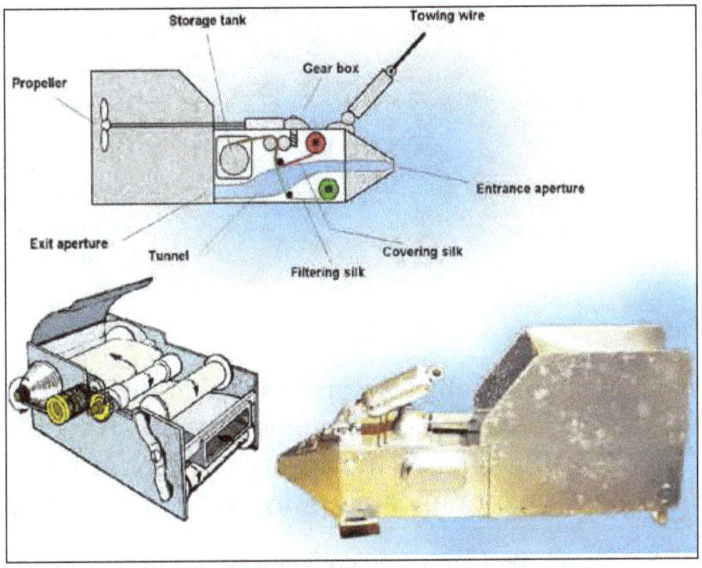

Continuous Plankton Recorder.

The continuous plankton recorder (CPR) survey is the largest multi-decadal plankton monitoring programme in the world. Since the prototype was deployed, until 2004, over 4 000 000 miles of towing have resulted in the analysis of nearly 200 000 samples and the routine identification of over 400 species/groups of plankton, used to study biogeography, biodiversity, seasonal and interannual variation, long-term trends, and exceptional events. Another and notably CPR most important aspect is the ability to collect hundreds of samples throughout an ocean basin because it is can be towed behind ships of opportunity (SOOPs) at their conventional operating speeds of 15–20 knots.

Recently, at the Woods Hole Oceanographic Institution (WHOI), was designed and realized a Bio-acoustic system, the BIOMAPER II. In its original conception BIOMAPER II was designed primarily for acoustic monitoring of plankton and includes both up- and down-looking acoustic transducers of different frequencies, as well as a suite of conventional environmental sensors (including conductivity, temperature, pressure, chlorophyll fluorescence and beam transmission). The upgraded vehicle has been integrated with a pair of dual path absorption and attenuation meters, one for whole water and the other for a filtered fraction (0.2 Pm), and two spectral radiometers for measuring down welling irradiance and upwelling radiance. The BIOMAPER II is particularly well suited to assessment of apparent optical properties during towed operation because the vehicle is designed to maintain a horizontal attitude regardless of flight pattern.

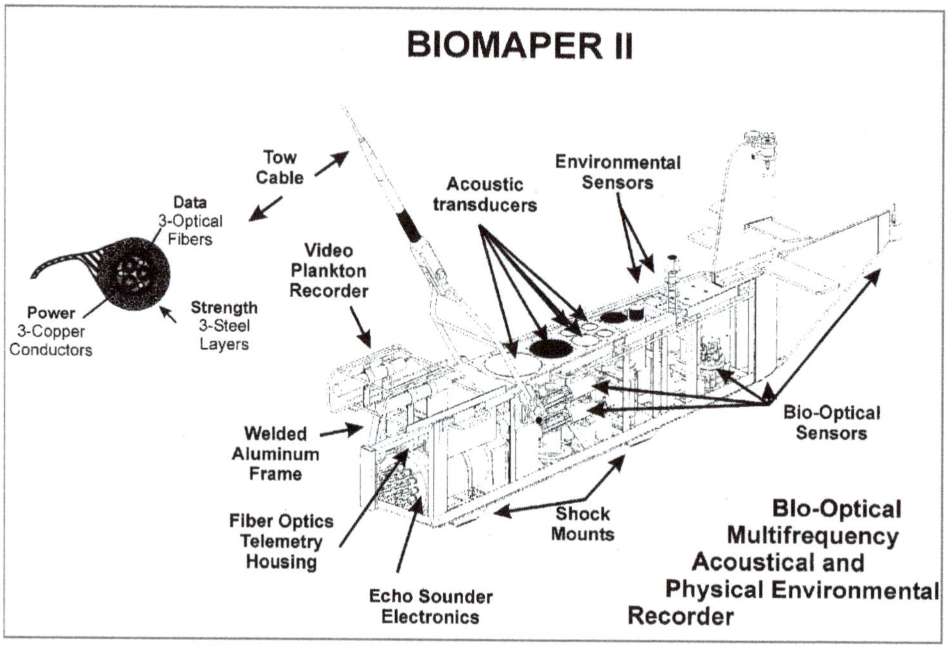

BIOMAPER II.

During the 1970s the firsts undulating vehicles were realized at the Bedford Institute of Oceanography (BIO), Canada and the Institute of Oceanographic Sciences (IOS of National Oceanography Centre, Southampton, UK). The BIO Batfish was a variable-depth towed body equipped with a CTD and various other sensors used for rapid underway profiling of the upper layers of ocean.

Its evolution is the SeaSoar developed by Chelsea Technologies Group from the original design of IOS. It is a versatile towed undulating vehicle used to deploy a wide range of oceanographic monitoring equipment. Its typical instrumentation payload is a combination of: CTD, Fluorimeter, Transmissometer, Turbidity, Bioluminescence, Irradiance meter, Nitrate/Nitrate sensor, Plankton Sampler, and SeaWifs bands sensors.

SeaSoar

In 1990 the italian underwater undulating vehicle SARAGO was planned at ISMES Laboratories in collaboration with University of Tor Vergata, Rome, Italy. Its main objective was to provide quasi-synoptic measurements of ecological and physic-chemical variables in marine environment. Its main application was the basic scientific research, the environmental monitoring and the assessment of biological resources. Therefore it was created a modular system capable to support different instruments and to expand its own working operating capabilities.

SARAGO characteristic payload consists of a SBE 19 CTD for physical variables and a pressure transducer to control depth separately from CTD sensors and a double impulse fluorometer called Primprod 1.11, created in collaboration with the Biophysics Institute of Moscow University. With this configuration, the SARAGO was used in the project PRISMA2 CNR from 1995 to 1998, to continuous measure photosynthetic efficiency and PAR, and to compute high resolution estimations of phytoplankton primary production.

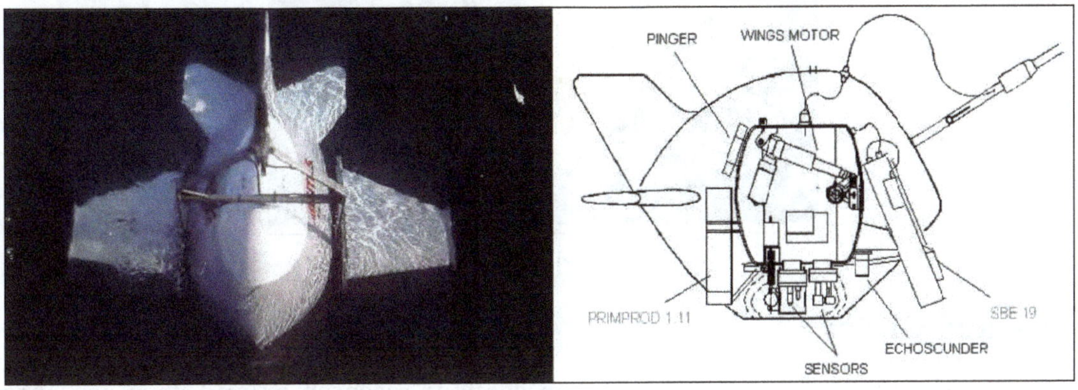

SARAGO.

SARAGO paths are checked by means of a pressure transducer for the wings and Sonar. The pressure transducer allows controlling depth separately from CTD sensors. The sonar works both as a path controller (in case SARAGO is programmed to navigate at constant distance from the bottom) and as a bottom alarm.

A different approach was taken by the researcher of Institute for Marine Environmental Research (IMER), Plymouth, UK. They targeted ships of opportunity with the robust and compact Undulating Oceanographic Recorder. Unlike its predecessors the UOR did not require a multi core tow cable all power being generated by an onboard impeller/alternator thus removing the need for a dedicated research vessel.

The Undulating Oceanographic Recorder UOR Mark 2 is a self-contained, multirole, oceanographic sampler. It is independent of the vessel for any service and can undulate from research vessels and merchant ships at speeds up to 13.5 m. s^{-1} (26 knots) between the surface to depths of 55 m (at 4 m.s^{-1}) and 36 m (at 10 m.s^{-1}) with a preset undulation length between 800 and 4000 m.

It can record (internally on a miniature digital tape recorder, resolution 0.1% full scale) measurements by sensors for temperature, salinity, depth, chlorophyll concentration, radiant energy, or other variables.

Since emerging oceanographic programmes demanded large synoptic data sets over extended periods of time with inherent reliability and high data recovery, so new towed vehicles are required to accommodate larger payloads, be highly versatile yet meet or exceed the performance of existing vehicles. Improved production techniques should also reduce the manufacturing cost.

One example is represented by the ScanFish MK II of Graduate School of Oceanography, University of Rhode Island. ScanFish is a towed undulating vehicle system designed for collecting profile data of the water column in oceanographic, bathymetric and environmental monitoring applications at either fixed depth or to pre-programmed undulating flight path. The ScanFish MK II is an active unit, which generates an up or down force in order to position the tow-fish in the water column. This instrument may carry on a number of sensors measuring parameters like conductivity (salinity), temperature, pressure, fluorescence, photosynthetically active radiation (PAR) and other parameters.

ScanFish.

ROV, AUV and Gliders

The technology of underwater vehicles focus on seafloor research or/and deep sea observation using human occupied submersibles (i.e. Alvin) or remotely operated vehicles (ROVs), and most recently, autonomous underwater vehicles (AUV). Despite AUVs represent the ultimate advances of oceanographic platforms they are very expensive due to involved technologies and deployment systems.

GLIDERS, are the most recently developed class of underwater vehicles. Due to their autonomous capacity to perform saw-tooth trajectories driven by changes of buoyancy, they can be considering as low cost AUVs. The first modern deep Human occupied vehicles, was the William Beebe bathysphere able to dive down to 923 m depth.

Realized in 1934, the bathysphere was a small spherical shell made of cast iron, 135 cm in inside diameter designed for two observers. The second generation of deep submersibles began in 1964 with Alvin. It was funded by the US Navy under the guidance of the Woods Hole Oceanographic Institution (WHOI). Alvin was a three-person, self-propelling, capsule-like submarine about 8 meters long that could dive up to 3.600 meters. It was equipped with underwater lights, cameras, a television system, and a mechanical manipulator.

In 1975, scientists of Project FAMOUS (French-American Mid-Ocean Undersea Study) used Alvin to dive on the Mid-Atlantic Ridge for the first direct observation of seafloor spreading. Alvin has allowed the discovering of the hot springs and associated sea life that occur along the East Pacific Rise. Moreover, it provided early glimpses of the Titanic wreck site in 1985. Since Alvin, other manned submersibles have been used successfully to explore the ocean floor.

Remotely operated vehicles (ROVs) are underwater platforms remotely controlled from the surface by a cable that provides power and control communication to the vehicle. ROVs can be divided into three main types of vehicle:

- Free-swimming tethered vehicles have thrusters that allow manoeuvring along the three axes and provide visual feedback through onboard video cameras. Generally their use is for mid-water or bottom observation or intervention.

- Bottom-crawling tethered vehicles move with wheels and can only manoeuvre on the bottom. Visual feedback is provided by onboard video cameras. Their mainly use is for cable or pipeline works.

- Towed vehicles are carried forward by the surface ship's motion, and are manoeuvred up and down by the surface-mounted winch. Towed vehicles usually carry sonar, cameras, and sometimes sample equipment.

ROVs have been developed in the 1950s and have been used worldwide since the 1970s and 1980s to explore previously inaccessible underwater supporting scientific missions. Autonomous underwater vehicles (AUVs) are untethered mobile platforms, computer controlled, used for survey operations by ocean scientists, marine industry, and the military. Their main feature is the completely free/or very little human operator interaction while operating. This characteristic allows AUVs to access otherwise-inaccessible regions as for example under Arctic and Antarctic ice.

Despite the off-shore oil industry represents the main field of use of AUV platforms, since 1990s the oceanographic scientific community has adopted these autonomous vehicles for oceanographic investigation as they can provide, for example, the rapid acquisition of very distributed data sets. Today a wide range of AUV is available for commercial use as well as for oceanographic surveys, and a large number of companies develop subsystems and sensors to install onboard AUVs.

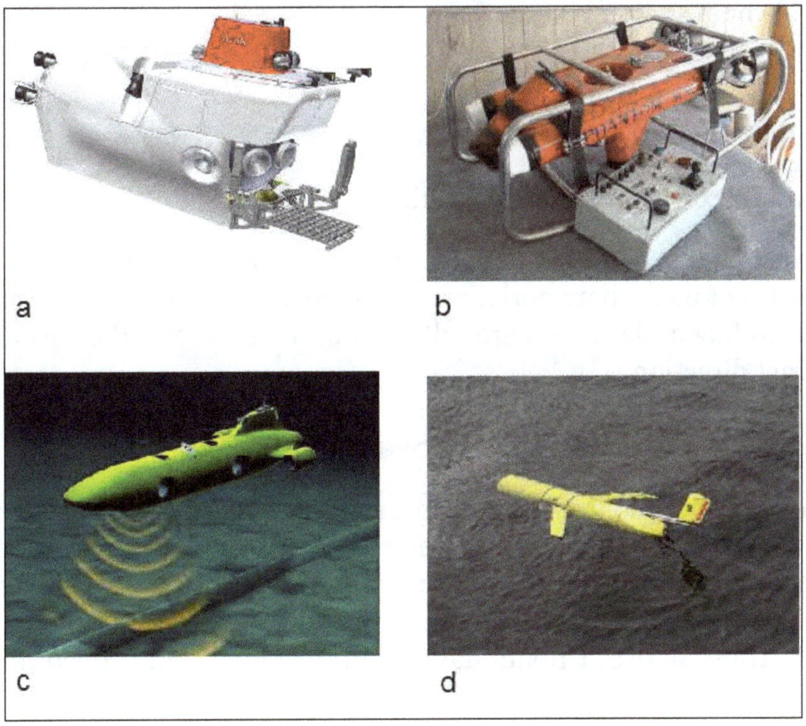

(a). Alvin (b). ROV (c). AUV (d). GLIDERS.

AUVs are highly integrated devices, containing a variety of mechanical, electrical, and software subsystems that can be summarized as follow:

- Hardware and software components in order to program specific tasks.

- Energy storage to provide power.

- Propulsion system.
- Stability control system.
- Measurement instruments.
- Communication devices.
- Locating devices.
- Emergency recovery system.

Gliders have been developed specifically to better investigate the main characteristics of mesoscale eddies, fronts, and boundary currents. The salient characteristic of gliders can be found in the realization cost that corresponds to about few days of research vessel operation while they can be operated for a few months.

Three operational gliders, Slocum (Webb Research Corp.), Spray (Scripps Institution of Oceanography and Woods Hole Oceanographic Institution) and Seaglider (University of Washington) were realized within the Autonomous Oceanographic Sampling Network program set up by the US Office of Naval Research (ONR) in 1995. The four key technical elements of gliders are:

- Small reliable buoyancy engines.
- Low-power computer microprocessors.
- Gps navigation system.
- Low-power duplex satellite communication.

The three operational gliders are descendent floats and differ from drifter and lagrangian profilers essentially because of their characteristics: the wings that allow forward motion mainly while diving. During this gliding cycle gliders are fully autonomous, and at the surface they use GPS to determine next target direction.

Lagrangian Devices, Expendable Probes and Drifter

Traditional methods for oceanographic measures and observations are carried out through ad hoc surveys in order to collect data of specific areas, with the consequently limitation of time and costs. To overcome this limitation, world oceanographic community, has developed different programs and methodologies. Starting in the 1970s, a great number of surface drifters and subsurface neutrally buoyant floats have been developed, improved, and tracked in the ocean.

Drifters are able to provide worldwide maps of the surface and subsurface water velocity at a few depths. Lagrangian profiling buoys like WOCE floats and their successors Argo are measuring and reporting in real time the temperature and salinity structure of the upper 2km of the ocean. Drifters and floats provide fundamental surface and sub-surface data allowing the description of World Ocean's temperature and salinity structure, ocean circulation and its time variability.

The basic elements of a lagrangian drifter designed, for example, to track mean currents at a fixed depth beneath the ocean surface are: the drogue, the surface float and the connecting tether. The drogue is the draging element that locks the drifter to a water parcel. The surface float contains the telemetry system, antenna, batteries, and sensors. The drifter position can be calculated by an installed GPS receiver. Data and position can be transmitted by Argos or Iridium systems to the data server.

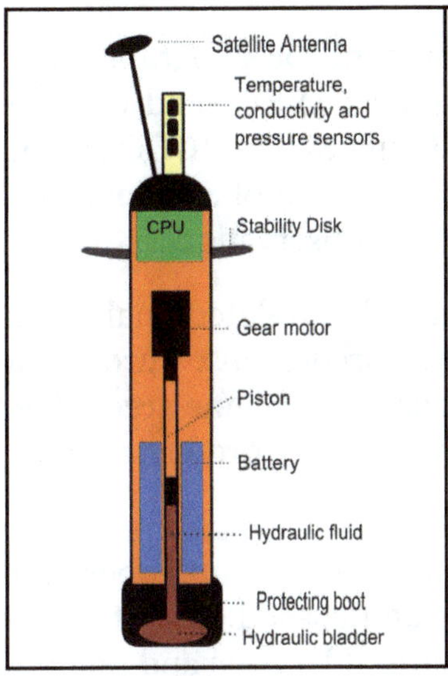

Argo.

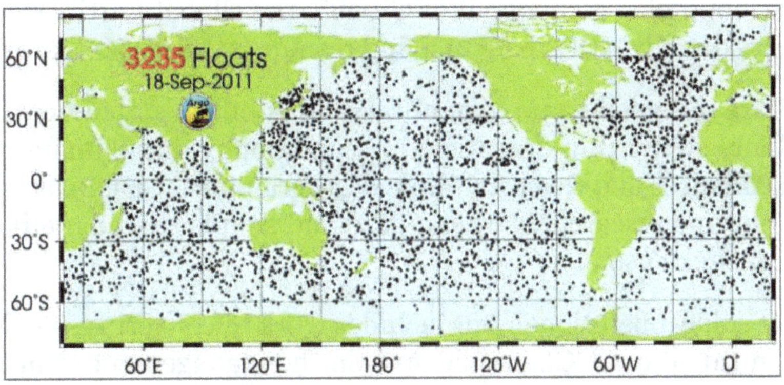

Argo distribution map.

Autonomous WOCE and Argo float, usually drifts submerged for a few weeks. Periodically, when it rise up to the sea surface, transmit data and position by the Argos satellite system. After around a day drifting on the surface, the float re-submerges to its operative depth, in the upper kilometer of the ocean, and continues to drift for another few weeks. Around 100 round trips are possible over a lifetime up to 6 years.

Despite Argo floats can autonomously measure physical, chemical and some biological parameters, in practice this potential capability is limited by cost and power supply. Recently, expendable probes like expendable Bathy Termographs (XBT), have assumed particular relevance for the measure of physical variables. Their main future is the possibility to be launched directly from a moving ship, saving time and costs.

Despite all technological developments, the challenge of oceanography still remains the synopsis of data acquisition in order to observe high spatial and temporal variability phenomena such as oceans primary production. This variable is fundamental as it is strictly related with oceans water characteristics (physical, chemical and biological) and it regulated by ocean dynamics (current and water mass properties).

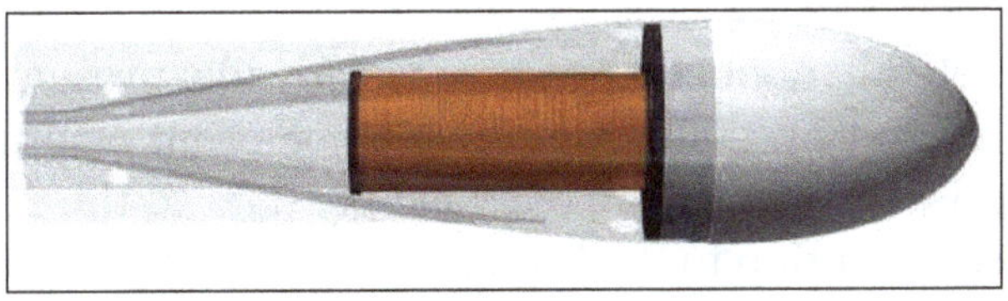

XBT.

The JCOMM Ship Observations Team (SOT) Programmes: VOS and SOOP

From the origin of the climate studies, oceanographic measures have been acquired by ships, the only available platforms to survey the oceans. Although there are now several other platforms (i.e. satellites, drifting buoys, floatings and radar) ships still play a very important role providing sea truth for the calibration of satellite observations and because of they allow measurements not yet obtainable by other means, such as air temperature and dew point.

The World Meteorological Organization (WMO) within Intergovernmental Oceanographic Commission (IOC) of UNESCO, the United Nations Environment Program (UNEP), and the International Council for Science (ICSU) sponsor the Global Ocean Observing System (GOOS) that is the oceanographic component of GEOSS, the Global Earth Observing System of Systems. WMO members directly participate in living resources, the GOOS, providing in observations.

GOOS are designed to provide observations of the global ocean (including related analysis and modelling, supporting operational oceanography and climate change predictions. Through Regional Alliances and dedicated linked programmes, GOOS allow all data to become accessible to the public and researchers by standard and simple shared methods to produce products useful to a wide range of users.

It is not solely operational, but includes work to convert research understanding into operational tools. In this way GOOS should sustain the management of marine and coastal ecosystems and resources, supporting human activities like mitigate damage from natural hazards and pollution, protect life and property on coasts and at sea and suggest scientific research activities. Since 1999, the marine activities of WMO, as well as those of IOC, have been coordinated by the Joint Technical Commission for Oceanography and Marine Meteorology.

The creation of this Joint Technical Commission results from a general recognition that worldwide improvements in coordination and efficiency may be achieved by combining the expertise and technological capabilities of WMO and IOC.

The Observations Program Area of JCOMM is primarily responsible for the development, coordination and maintenance of moored buoy, drifting buoy, ship-based and space-based observational networks and related telecommunications facilities. In this framework it was created a Ship Observations Team (SOT) which coordinates two programs: the Voluntary Observing Ship Scheme (VOS) and the Ship of Opportunity Program (SOOP).

The WMO Voluntary Observing Ships (VOS) Program

Developed about 150 years ago, VOS is the international scheme for taking and transmitting meteorological observations by ships recruited by National Meteorological Services (NMSs), comprising member countries of the World Meteorological Organization (WMO).

With a fleet strength, estimated in about 4000 ships worldwide, presently the VOS supports measures useful to comprehend and predict extreme weather events, climate variability, and long-term climate changes. In this way, providing increasingly data to the global climate studies, VOS makes a highly important contribution to the Global Observing System (GOS) of the World Weather Watch (WWW).

The Ship-of-Opportunity Program (SOOP)

The SOOP is based on the possibility to use merchant ships that join to the programme, in routinely strategic shipping routes. At predetermined sampling intervals, Expendable Bathythermographs (XBTs) are launched to acquire temperature profiles in the open ocean in order to be assimilated into operational ocean models.

SOOP plays a relevant role as it serves as a platform for other observational programmes, communicating closely with the scientific community. In this way, SOOP assumes a big relevance to seasonal and interannual climate prediction.

The programme is managed by the SOOP Implementation Panel (SOOPIP). Along the strategic shipping routes, SOOPIP identifies and coordinates the measuring necessities in terms of type of instruments, their use modality and deployment. So, thanks to the use of different instruments such as XBTs, TSGs, CPR, it is possible to obtain long measuring series of physical, chemical and biological parameters.

Moreover SOOP looks at new instrumental technological development and coordinates the exchange of technical information; in particular: functionality, reliability and accuracy, and recommended practices about relevant oceanographic equipment and expendables. The main important aspect of VOS and SOOP programs deals with the possibility of collect data in a cost effective way.

VOS operates without ship cost, whilst the communication charges, the observing equipment's and the consumables are furnished by the National Meteorological Services. Within these programs (VOS and SOOP), many international initiatives are presently carried out in order to develop new observing devices and systems, suitable to be utilized from as many vessels as possible.

New expendable probes and devices (low cost and user friendly) represent the ultimate effort of oceanographic technological development in order to collect even more accurate measurements and to enhance the range of variables to be measured in automatic way, particularly for the biological ones.

Remote Sensing Platforms

Data from satellites allow us understanding and monitoring ocean processes at large spatial and temporal scale. Also if the same sensors payload equips airborne remote sensing, this platform is otherwise dedicated to the surface analysis of meso-scale processes, particularly to that of shallow waters and coastal environment.

Even if remote sensing can provide a unique synoptic view of phenomena such as chlorophyll concentration and sea surface temperature, however the measurement validation requires constant scrutiny and much sea-truth data as possible. For this reason (and also to develop ecosystem forecasting models) the integration between field sampling and remote sensing technique should be improved. Actually this is a primary necessity also because of still a lack both of basin wide and of coastal ones, operational, multidisciplinary, in situ observing system.

In fact today, thanks to new satellites capabilities, there is the possibility of sampling on smaller scales to describe relevant phenomena, especially in the coastal zone where scales are considerably smaller than those in the open ocean. To comprehend the complexity of physical and biological processes, the human activities impacts, the climate changes, and every process involving seas and oceans, it is not possible to use a single approach. To face this problem, the best strategy seems to be the integration of data derived from different platforms.

For all these reasons, the oceanographic investigation needs more and more new instruments and sensors with well-defined measurement characteristics also in order to calibrate the synoptic data providing by remote sensing observations. Whilst the monitoring of physical variables is continuously improving, the biological ones still lack and furthermore they have to be observed in situ.

There are several empirical algorithms commonly used to assess primary production in the euphotic zone, based on data provided by satellites Coastal Zone Colour Scanner. These models are parameterized both on water visible radiance (obtained by means of satellites), and other simple variables, and also on the knowledge of photo adaptive parameters, such as maximum photosynthesis which can be obtained only by in situ sampling.

Requirements of Oceanographic Observations

One of the most important problems of operative and operational oceanography is related to the availability, quality and accessibility of the data. To meet these needs it is necessary to acquire more data than possible, but, as already mentioned, data collection is often limited by survey costs. Moreover data distribution is often discontinuous and it does not allow a synoptic understanding of many dynamic processes.

The ships of opportunity are fundamental, since they integrate the ocean surface temperature (from satellites) with water column data. So they contribute to the Marine Forecasting System capability of providing near real time analysis of the oceans.

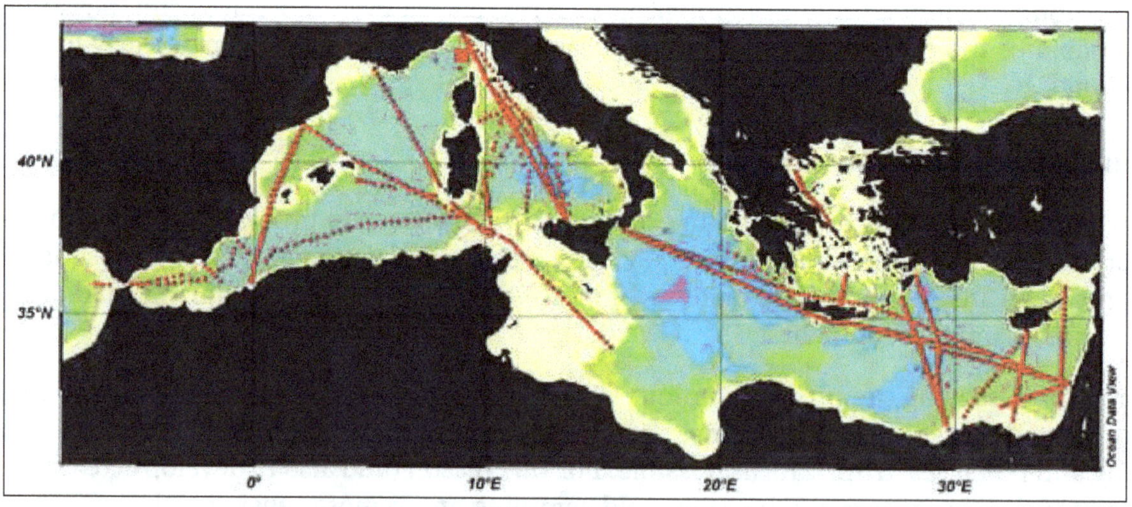

Mean Mediterranean VOS acquisitions (MFSTEP Project).

Within the international projects knowledge necessities, it is fundamental the development of innovative technologies according to requirements of high performances, data quality and low-cost. Even if data provided by models, by high resolution remote sensing and by the new autonomous sampling platforms are increasing temporal and spatial sampling capabilities, however, there is still a lack of operational, multidisciplinary, in situ observing systems. This fact is particularly related to the biological variables that need to be observed in situ more than the physical once. Especially in the mid-high latitudes, a complete upper layer observation of the water column is needed, because of the typical distribution of phytoplankton's biomass.

Traditional methods (water sampling, storage and laboratory analysis) are too expensive and do not allow to have enough measures to describe variability of marine natural phenomena with sufficient temporal and spatial detail.

Therefore, bio-optical measures, used to study the main environmental characteristics, such as phytoplankton biomass, CDOM (Coloured Dissolved Organic matter), turbidity, can be integrated in new technological developments to give continuous measures.

In such a context our aim was to answer to all the above issues, developing new user friendly technologies, based on low cost materials and suitable to different configurations (Moored, stand alone, expendable, continuous and towed). In order to face the necessity of the world oceanographic observations, we identified two main development necessities:

- A new low cost flexible modular instrument, to be used like profiler, expendable, stand alone and instrumental payload.
- A towed vehicle able to be used both in coastal and open oceans, capable to host the new low cost instrument.

Low Cost Modular Instrument

Expendable probes (see before XBT), represent an approach to ocean measurements in which the high accuracy of measurements may be sacrificed considering lower costs and operational expediency. Relating to operational oceanography, this kind of technology derives from the needs of adequate spatial sampling on timescales commensurate with temporal variability.

These probes can provide measures of temperature quickly as they can be used by a ship moving up to 20-30 knots. This technique also provides standard results and it became the central component of programmes such as the Global Ocean Observing System (GOOS).

Current expendable probe capabilities include also the measurements of sound speed, conductivity, ocean current and (most recently) optical irradiance and suspended particle concentration. Such expendable measurements could be fundamental for satellite data calibrations and for programmes such as Marine Forecasting System (MFS).

With support from GOOS, GEO, POGO, IOCCG and Plymouth Marine Laboratory, the Chlorophyll Globally Integrated Network (ChloroGIN) was set up in order to provide a network of researchers with the aim of integrating in situ time series of chlorophyll measurements with satellite ocean colour-based observations. From these integrated measurements it is possible to provide maps of ocean chlorophyll and sea surface temperature, as indicator of the state of the ecosystem, and measures of light penetration to calculate primary production.

In addition to the prototypes made by Marcelli et al. during the MFS-TEP project, at now there are no expendable probes for the measurement of chlorophyll and other bio-optical variables.

This last feature is very important as it may concern in situ measurements of water optical property dealing directly with phytoplankton biomass and Coloured Dissolved Organic Matter (CDOM).

The measure methods currently available consist in fact in expensive instruments, or they require expensive methods (sampling and laboratory analysis). In this field, the maximum effort should be made to develop innovative tools looking to: economist, flexibility and with a sufficient level of precision and accuracy. They must be modular to be adapted to measure the widest number of variables and they must be flexible to be used by the largest possible number of measurement platforms, and also as expendable.

Towed Vehicle

In order to increase our understanding of ocean processes it is necessary to have the most comprehensive and synoptic data sets. Significant advances have been made in remote sensing from both satellite and aircraft, which nevertheless increase the requirement of in situ data for sea truth. Methods of traditional oceanographic research and marine environment monitoring are limited to the time scale of the observer and therefore they cannot carry out synoptic observations.

The large coverage collection, of long term data sets or high resolution field studies, should require high costs. The problem of monitoring of physical, chemical and biological parameters of the oceans, both for operational and forecasting remains one of the most important issues that ocean science community is facing today.

It is commonly accepted that the most cost effective method for spatially and temporally monitoring is represented by the use of both towed instrumentation and expendable low-cost probes from ships of opportunity.

Towed vehicles, in use since the 1930s, in recent years have been advanced significantly becoming extremely robust and high reliable. Typical payloads are CTDs, which can be interfaced with other sensors such as fluorometers, transmissometers, pH, dissolved oxygen, and other optical sensors dedicated to sea truth.

Undulating vehicle main feature is the capability to make continuous measurements along water column by moving ship, with a considerable operative time saving. Furthermore continuous measurements can resolve fine structures of high variability phenomena such as phytoplankton patchiness.

Distribution of phytoplankton biomass is characterized by a great variability. Its assessment is a fundamental issue since it modulates the carbon cycle. While primary production measurements are expensive for time and costs, phytoplankton biomass can be detected by in-vivo chlorophyll a measurements.

The possibility of using a continuous profiling probe, with an active fluorescence measurement, is very important in real time phytoplankton's study; it is the best way to follow the variability of sea productivity. In fact, because of the high time and space variability of phytoplankton, due to its capability to answer in a relatively short time to ecological variations in its environment and because of its characteristic patchiness, there isn't a precise quantitative estimation of the biomass present in the ocean.

Despite this fundamental aspect, the use of such towed vehicle often needs dedicated surveys or ships with ad hoc deploying and towing systems. However, towing vehicle will continue to have a key role if deployed from ships of opportunity looking to oceanic processes and it will become fundamental for the study of coastal high variability processes.

Physical Oceanography

The Physical Oceanography is an essential part of oceanographic analysis. It is the study of physical conditions that are prevailing in the seas and oceans. It deals with all large scale physical processes and their effects that are happening within the oceans. The processes are very dynamic in nature. They involve the water masses which have heterogeneous proportions and dimension. The properties of water masses also vary with space and time. Physical oceanography considers all these aspects in projecting the oceans. This branch has grown due to historical oceanographic explorations and expeditions carried out by several scholars from almost all parts of the world. Very essential aspects of physical oceanography are to be understood first.

Essential Aspects of Physical Oceanography

The very essential parts of the subject of physical oceanography include the following aspects. Physical oceanography invariably deals with the aspects like the Physical Setting of the seas and oceans, Historical oceanographic explorations, Instruments used in oceanographic studies, Atmospheric Influences of oceanic water masses, The Oceanic Heat Budget, Temperature, Salinity, and Density of oceanic masses, The Equations of Motion and Equations of Motion With Viscosity, Air-sea interface and response of the Upper Ocean to Winds, Geostrophic Currents and Wind Driven Ocean Circulation, Vorticity and Deep Circulation in the Ocean, Equatorial Processes and their effects, Ocean Waves and Tides, Numerical Models, Coastal Processes and Ocean induced hazards. It would be very appropriate to understand the evolution of the subject of physical oceanography by knowing it through some of its historical explorations, made by the people.

A lot of oceanographic expeditions and explorations have been conducted by several groups of people under the support of ancient and modern kings and kingdoms. All those investigations have helped people over the years to understand the nature of oceans to a great extent. It is progression with many innovations and discoveries, even today. Still there are a lot of things found inside the deep oceans. The unexplored regions are more than the explored domains. Oceans offer a lot of marine resources, fisheries, mineral deposits, food resources and energy resources. These are much useful to the human population of the whole globe to survive. It is necessary to understand the integrated aspects of the subject of oceanography with all the available scientific facts, principles, findings and observations.

The Physical Setting

The oceans are very thin layer of water masses when we consider them on a global scale. The vertical scale is very small when compared to the horizontal scale. Hence, for all graphical representation, the vertical scale is to be exaggerated. The ratio of depth to width of the ocean basins is also very small. This ratio is very important for understanding the ocean currents. The Vertical velocities must be much smaller than the horizontal velocities. Very small amounts of vertical velocities may have a great influence on the water turbulence in these bodies.

World's Oceans and Seas

The Earth has only one large marine water mass. It is distributed in the forms of seas, oceans, gulfs, straits, basins and bays. The total area covered by these water masses is 361 million sq.km. The total volume of this mass is 1370 million cu.km. It is higher than the volume of land above sea level which is only 130 million cu.km. The total mass of the oceans is 1.5 x 10^18 tons, which is 0.25 % mass of the earth. There are five major oceans on the earth as the Pacific, the Atlantic, the Indian, the Arctic and the Antarctic Oceans. All these water masses cover about 70.8% of the surface of the earth.

Dimensions of the Ocean

The dimensions of oceans may range from 1500 Km to 13000 km along latitude or longitude. The minimum width 1500 km is seen in the Atlantic ocean along east-west direction and it extends to more than 13000 Km along its North-South extension. The surface area of the Pacific Ocean is 181.34 million Sq.K m. Its average depth 4282 m. The surface area of the Atlantic Ocean is 106.57 million Sq.Km. Its average depth is 3926m. The surface area of the Indian Ocean is 74.12 million Sq.Km. Its average depth is 3963m. The overall average depth of oceans is 3800m. If we look at the depth regions throughout the oceans, we can see that majority of the area is occupied by the depth zone between 2000 and 6000m. But the deepest zones beyond 6000m depth are noticed only in a few places in the west and northwest Pacific, mid-west Atlantic and a few isolated spots of the Indian and other oceans.

Sea-Floor Features

Bathymetry is the mapping of ocean-water column. It provides depth-wise relief features. The depths of the oceans are measured using acoustic echo-sounders housed on ships and altimeters housed in Satellites. Bathymetrically, the relief of the oceans is divided into 3 distinctive regions as Continental Shelf, Continental Slope and the deep ocean basins. All these three zones have unique physico-chemical and biological conditions. They are also characterized by distinct geological features and earth's processes. The morphology and Relief of the Sea-Floor include a wide variety of features. They include the submarine canyons, deep sea fans, abyssal hills, abyssal plains, mid-ocean ridges, deep sea trenches, seamounts, rift-valleys, ocean fractures, oceanic troughs and island arcs.

Crustal Plates

The earth's crust is broken into various lithospheric plates that move relative to each other. There are convergent and divergent plate movements. Both of these movements are responsible to create new ocean floors at one side and subducting the ocean floors ate the other side. Altogether, the relief of the oceans is under dynamic changes due to these continuing tectonic movements. Volcanic eruptions, earthquakes, rift-valleys, rising of ridges, crushing of continental margin materials, and tsunamis are expected during these processes.

Study of Sun's Radiation

All dynamic processes in the oceans are directly or indirectly controlled by the Sun's radiant energy and the atmosphere. The dominant mechanisms are the role of Sun light, evaporation of sea water, infrared emissions from the sea surface and the cooling (or) warming of ocean waters. The atmospheric circulations are controlled by the oceans and the ocean water circulations are controlled by the atmosphere. Oceans are also the dominant sources and sinks of heat. Oceans can absorb, re-emit, store and transfer heat very effectively by its constituent water masses.

Solar Heating

Solar heating is an uneven energy source. This uneven heating controls the strength and direction of winds in the atmosphere. The loss and gain of heat by the oceans produce winds in the atmosphere. Evaporation and condensation are two more mechanisms enacted by the heat in oceans and cooling in the atmosphere. About 50% of the solar energy reaching the Planet Earth is absorbed by the ocean and by the land. One-fifth of the solar energy absorbed by the ocean is released back to the atmosphere, through evaporation and infra-red radiation. The balance is then transferred into sea and is used for producing currents. The Heat Flux is the transfer of heat across the ocean and along the sea surface. The flux of heat on water also changes the density of water and its buoyancy. It is estimated that about 4000 joules of energy is required to heat 1.0 Kg of sea water by 1 °C.

Effects of Heat on Oceans

Earth gains heat at the top of the tropical atmosphere, and it loses heat at the top of the polar atmosphere. The Isolation is the term denoting the incoming solar radiation. It is determined by latitude, season, time of the day and cloudiness. The Polar Regions are heated very less than the tropics. North-South movement is called as meridonial heat transport. Areas in winter are heated less than the same area in summer. The heat in the morning time is less than the Noon hours. There is no insolation at night. All these have an impact on the temperature of water in seas and oceans.

Atmospheric Pressure Zones

Due to these, high and low atmospheric pressure zones are also created. Warm and cold fronts are generated out of these changes. The atmosphere within 100m above the sea surface is influenced by the turbulent drag of the wind. This layer of atmosphere is called as the boundary layer.

Wind Movements

Wind measurements, in seas and oceans, have been made for several centuries. The world's oldest wind reports were made as early as 1855. Wind speeds were understood through a convenient system of reference as proposed by Sir. F. Beaufort in 1806. This is called as the Beaufort Wind Scale. This scale is a graded level of denoting the force of the wind and its impact on water and on land. It helps to know the severity and intensity of the wind on the sea. The force level ranges from 0 to 12, representing calm wind to hurricane on water masses. Later, the conditions of the sea were also determined using the modified Beaufort wind Scale proposed by Kent and Taylor. Weather observations on ships including the Beaufort wind force are measured, normally four times a day, all over the world. They are done at 0 hours, 6 AM, 12 noon and 6 PM at local times.

Ocean Water Properties

The properties of sweater are essential aspects of physical oceanography. The sea water has certain physic-chemical properties which play a very significant role in the hydrodynamics and also on the marine life. The notable properties are the temperature, salinity, density and light penetration. Temperature and Salinity measurements with reference to depth are done using Bathythermography, Expendable Bathythermography and Nansen Bottles.

Salinity of Seawater

Salinity is the total amount of dissolved solids present in sea water. It is represented as gm per Kg of sea water. It is a dimensionless quantity. It ranges from 34.6 to 34.8 parts per thousand. It is also indirectly determined based on conductivity and chlorinity of waters. Sea water contains higher concentrations of all cations and anions. However, sodium and chloride are the dominant species present in seawater. Conductivity and Salinity are measured using conductivity meters. The mean salinity of ocean water is found to be 34.7.

Density of Seawater

The density of water at the sea surface is typically 1027 Kg/m^3. Changes in density generate vertical movement. Turbulence in the mixed layer mixes the heat downward. The flow of heat, evaporation, rainfall, river water inflow, freezing and melting of the sea ice, all control the distribution of temperature and salinity of the water in oceans. Changes in these two properties can increase (or) decrease the density of water at the surface. It also promotes convection. When the surface water sinks into the deeper ocean, it creates much pressure variations and generates ocean currents.

Temperature of Seawater

Sunlight heats the sea water, warms the surface layers, provides the energy for all phytoplanktons and helps in the navigation of many marine animals. Warmest ocean water exits near the equator and the coldest water exists near the poles. The mean temperature of ocean water is 3.5 °C. Below the mixed layer, the water temperature decreases rapidly with depth.

Light Penetration

Light penetration and absorption of light by oceanic waters are very important aspects for marine life to survive.

Fluid Mechanics and Ocean Dynamics

Fluid mechanics and ocean dynamics are very essentials aspects in oceanography. Ocean waters are subjected to internal and external forces. The fluid mechanics and hydrodynamics are the two major concepts underlying many processes and life support systems in marine environments. The aspects included in this component are Fluid mechanics, Newtonian mechanics, Conservation of Mass, Conservation of momentum, Angular momentum and Conservation of energy.

Conservation Laws

The other essential part of study in Physical oceanography is the Conservation of Mass and Salt. These help in deriving the continuity equation and for understanding the dynamics of mass and force. Many equations of Fluid motion are derived from the Conservation Laws. The approaches employed under these concepts are:

- Conservation of mass, which leads to the derivation of Continuity Equation.
- Conservation of Heat Energy which leads to the computation of Heat Budgets.
- Conservation of Mechanical Energy which leads to the formation of Wave Equation.
- Conservation of Momentum which leads to the use of Navier Stokes Momentum Equation.
- Conservation of Angular Momentum which leads to the use of Conservation of Vorticity.

Air-Sea Interface

Air-Sea Interface is a very unique zone on the earth. It is the zone which will have the maximum atmospheric influences. The atmospheric influences come due to Sun's radiant energy on one side and the wind driven water movements at the other side.

Forces Acting on Masses

The forces which are essential to understand the Physical Oceanographic processes are Gravity, Friction and Coriolis force. Forces are vector quantities. They have both magnitude and direction. Gravity is the dominant force. The weight of water in the ocean generates pressure. Due to solar lunar influences, tides, tidal currents and tidal mixing all happen in the oceanic waters, regularly. Buoyancy is another parameter to be borne in mind. It is the upward (or) downward force arising due to gravity. It acts on a parcel of water. Density plays a catalytic role for enacting the Buoyancy. Horizontal Pressure Gradients are also created due to the varying weight of water in different parts of the oceans. Friction of wind blowing over the waves, force of water movement or air movement over water may create the frictional forces. Wind stress is also created due to this motion. In addition to these, Pseudo-forces and Coriolis force are also acting on oceanic waters. Coriolis force results from motion in a rotating co-ordinate system.

Ocean Water Circulation

Ocean water circulation also includes origin, and movements of Waves and Currents. The following are the types of circulation seen in the oceans:

- The General water circulation: This is a permanent, time-averaged circulation.
- The Deep water circulation: This is driven by mixing in the deep ocean.
- The wind-driven circulation: It is caused by local winds on the surface of oceans.
- The Gyres: These are wind-driven cyclonic (or) anti-cyclonic currents of giant magnitude.
- The Boundary Currents: These are currents flowing parallel to the coasts.

- The Jet streams (or) Squirts: These are long narrow currents with dimension of few 100 Km perpendiculars to West Coasts.
- The mesa-scale Eddies: These are turbulent or Spinning flows, running for a few 100 Km.

Motion with Viscosity

The role of friction in fluid flows and the stability of the flows to small changes in Velocity (or) density are to be understood in Physical Oceanography. Viscosity is the tendency of a fluid to resist shear. The influence of viscosity, especially the molecular viscosity which comes due to striking of water with solid mass, is to be analyzed. The components of Stress at a point in a fluid having pressure include normal stress and other shear stresses.

Turbulence

Turbulence is an important aspect in oceanic waters. Molecular viscosity is also affected by turbulence. Turbulence in the oceans leads to mixing. Because the ocean has stable stratification, vertical displacement must work against the buoyancy force. Vertical mixing requires more energy than horizontal mixing. The vertical mixing is very important in oceans, as it brings up the deep waters in some regions upward.

Stability of Mass

In oceanic waters, stability is yet another parameter to be understood. Many kinds of instability occur in these water masses. The kinds of stability to be understood are, static stability, dynamic stability, and double –diffusion. The Static Stability is associated with change of density with depth. The Dynamic Stability is associated with velocity shear and Double–diffusion is associated with Salintiy and temperature gradients in the oceans.

Ekman Layer and Langmuir Currents

Steady winds blowing on the sea surface produce a thin, horizontal boundary layer called Ekman Layer. This layer may be of few hundred meters in thickness. Ekman Layer transports water at right angles to the wind direction and subsequently generates upwelling. This upwelling enhances biological productivity to feed the fishers and other marine life. There is also a Deep Ocean Ekman layer near the sea floor. A current that spiral around an axis parallel to the direction of the wind is called as Langmuir circulation. It is customary to think that ocean currents are driven mainly by the winds. Currents and ocean circulations are driven by not only the winds but also other factors and forces caused by site, location and space on the globe.

Geostrophic Currents

The role of geostrophic currents is yet another mechanism in physical oceanography. The theory for wind-driven geostrophic currents was studied by the scientists Sverdrup, Stommel and Munk between 1947 and 1951. Satellite altimetric observations of the oceanic topography can give the time-variant data of surface geostrophic currents. Hydrographic data are used to calculate the internal geostrophic currents in the ocean. Flow in the ocean that is independent of depth is called barotropic flow and the flow that depends on depth is called as baroclinic flow. Baroclinic flow can be determined by using hydrographic data.

Types of Ocean Currents

There are several types of ocean currents existing in the oceans. They are:

- Antarctic Circumpolar Current.
- Deep Ocean Density-driven Current.
- Western Boundary Currents:
    - Gulf Stream.
    - Kuroshio Current.
    - Labrador Current.
    - Oyashio Current.
    - Agulhas Current.
    - Brazil Current.
    - East Australia Current.
- Eastern Boundary Currents:
    - California Current.
    - Canary Current.
    - Peru Current.
    - Benguela Current.

Vorticity

Vorticity is the rotation of the fluid. There are two types of vorticity as planetary vorticity and Relative vorticity. The sum of these two is known as Absolute vorticity. Vorticity strongly influences ocean dynamics. The curl of the wind stress adds relative vorticity to central Gyres of each ocean basin. In the oceanic waters, rotations and the conservation of vorticity stongly influence flow over distances exceeding a few tones of kilometers.

Equatorial Currents

The Deep circulation in the ocean is also an aspect to be understood. The Equatorial currents moderate the air-sea interactions, through the phenomena known as El-Nino. It is an acronym for "the child Jesus". These are found to be counter currents existing along the Peruvian coasts. The east-west temperature gradient on the equator drives a zonal circulation in the atmosphere. This is called as Walker circulation. El Nino denotes a disruption of this entire equatorial system along the Pacific. It causes changing weather patterns around the globe.

El Nino

The term El Nino is related to the southern oscillation (ENSO). El Nino is a sea-level pressure anomaly in the eastern equatorial Pacific Region. The tropical and equatorial Pacific is a vast ocean. It is a remote area which is seldom visited by ships. El Nino causes the biggest changes in equatorial dynamics. During El Nino, trade winds weaken in the western Pacific. This generates the Kelvin Wave eastward along the equator. These are the largest sources of annual fluctuations in global weather patterns. Due to these drought occurs in the Indonesian area and Northern Australia. Similarly, floods occur in Western, tropical South America.

Waves, Tides and Coastal Processes

In physical oceanography, a detailed study of the ocean waves, tides and their impacts on the coastal zone is needed. Ocean waves are powerful systems shaping and reshaping the coastal zones. Waves have energy. When we look at the sea from above, we can see thousands of waves on the sea surface approaching the coast. These are undulations of the sea surface with a height of around a meter or few meters. Waves break if the water is sufficiently shallow. The broken waves pour water into the surf-zone, creating long-shore and rip currents.

Ocean Waves

Waves are the natural and outstanding phenomena of the oceans. Wave phenomena involve the transmission of energy and momentum by means of vibratory impulses. For ocean-surface waves, the direction of propagation is perpendicular to the wave crests in the positive direction. Ocean waves are produced by wind. The faster the wind, the longer the wind blows. The bigger the area over which the wind blows, the bigger the oceanic waves generated. Observations by mariners on ships and satellites using sensors and altimeters help to prepare the global charts and maps of wave heights. In addition to the wave heights and movements, the directions are also detected by the modern sensors.

Wave Properties

Each wave has a wave height which is the vertical distance between the bottom troughs to the top crest of the wave. The wave length is the distance between two prominent adjacent crests or troughs. The wavelength may range from 50 to 100 m. When we watch these waves for a few minutes, we can notice that the wave heights and wave lengths are not constant. They vary in space and time. Wave period is the time taken by two successive waves to pass through a fixed point. All offshore waves are generated by wind. It is also true that the sea level changes from hour to hour.

Tides

In addition to these, it is also necessary to understand the Coastal Processes and Tides. Sea level increases and decreases during a day. The slow rise or fall of the sea level is due to the tides. Tides have wavelengths of 1000s of km. They are generated by the slow, very small changes in gravity as influenced by the moon and motion of the sun.

Tsunamis

Tsunamis are yet other issues in oceans and coastal regions. The impacts of tsunamis are very devastating. The recent Tsunamis have made the whole world population to think very seriously about natural disasters. Tsunamis are low-frequency ocean waves generated by underwater earthquakes. They are called as seismic-sea-waves. Sudden motion of sea floor over distances of a hundred or more km generated Tsunamis with wave periods of 15-40 minutes. These may propagate at a speed of 180 m/s with a wave length of 130 km in water. Tsunami's may be 3.6 km deep and is not noticeable at the seal surface. After nearing the coast, the surge may rise up to 10 or more meters above the sea level. A Tsunami can travel thousands of kilometers and do serious damage. The first wave of a Tsunami is not likely to be the biggest.

Tidal Energy

Oceanic Tides produce strong currents in many parts of the ocean. Their speed may go up to 5 m/s in the coastal waters. Oceanic tides have a great energy. They modify the coastal geomorphic conditions very often. Shorelines of submergence and Shorelines of emergence are two important aspects related to these waves and tides. Coastal erosion, coastal sedimentation and tidal inflow are all the natural effects of these processes.

Chemical Oceanography

Chemical oceanography is the study of everything about the chemistry of the ocean based on the distribution and dynamics of elements, isotopes, atoms and molecules. This ranges from fundamental physical, thermodynamic and kinetic chemistry to two-way interactions of ocean chemistry with biological, geological and physical processes. It encompasses both inorganic and organic chemistry, and includes studies of atmospheric and terrestrial processes as well. Chemical oceanography includes processes that occur on a wide range of spatial scales; from global to regional to local to microscopic dimensions, and temporal scales; from geological epochs to glacial-interglacial to millennial, decadal, interannual, seasonal, diurnal and all the way to microsecond time scales. The field by its own nature is very much an interdisciplinary field.

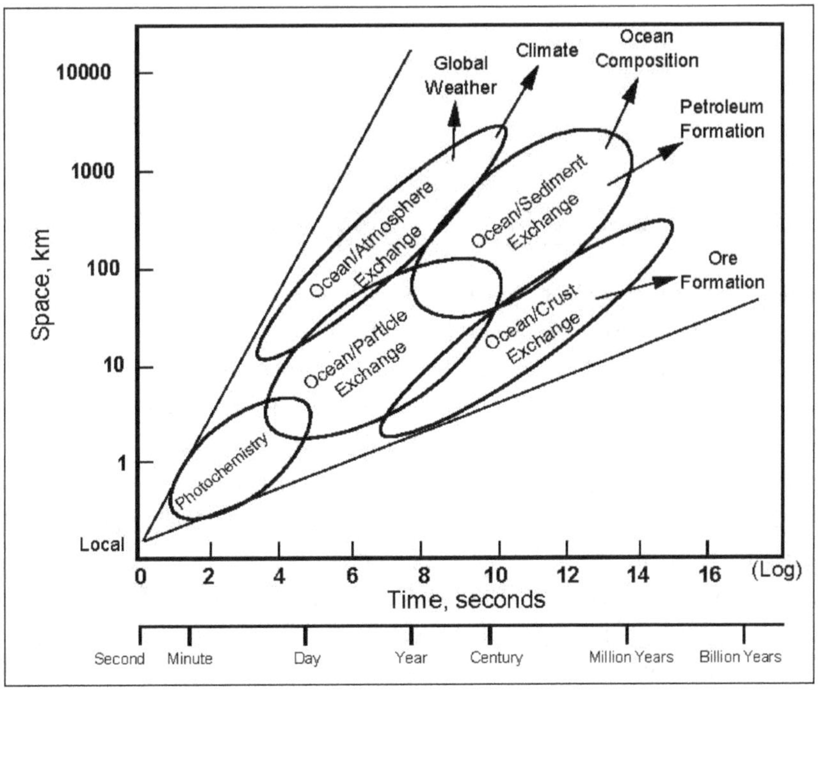

The advantages of the chemical perspective include:

- Huge information potential due to large number of elements (93), isotopes (260), naturally occurring radioisotopes (78) and compounds (innumerable) present in the ocean.
- Chemical measurements in the ocean are highly representative, reproducible and predictable (statistically meaningful). One drop of water is about 1/20th of a milliliter or 0.05g, this is 2.8×10^{-3} moles or 1.7×10^{21} molecules.
- Quantitative treatments are possible (stoichiometries, balances, predictions of reaction rates and extents).
- We can learn about processes from chemical changes. Seawater composition integrates multiple previous events, this is important because most of the ocean is inaccessible to direct observation.

We are all aware of the CO_2 and other greenhouse gases increase in the atmosphere since the beginning of the industrial revolution. This is seen in the classic data from C.D. Keeling showing the seasonal oscillations and the steady annual increase of CO_2 at the Mauna Loa Observatory. Most experts conclude that we are already witnessing the impact of this as global warming, and the signal is expected to become increasingly more pronounced.

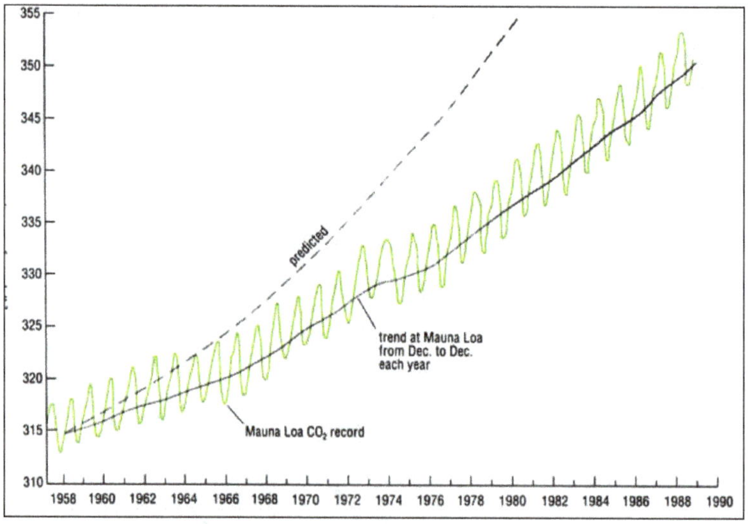

The inventory of dissolved inorganic C in the oceans is 50-60 times greater than that in the atmosphere, so a small perturbation of the ocean carbon cycle can result in a substantial change in the concentration of CO_2 in the atmosphere. The ocean carbon cycle influences atmospheric CO_2 via changes in the net air-sea CO_2 flux that are driven by differences in the partial pressure of CO_2 between the surface ocean and atmosphere. This exchange process is dominated by two interdependent "carbon pumps" that deplete the surface ocean of total CO^2 relative to deep water. Because the solubility of CO_2 increases with decreasing temperature, the SOLUBILITY PUMP transfers CO_2 to the deep sea during formation of cold deep water at high latitudes. This is a link of the ocean carbon cycle to physical processes (circulation). At the same time the BIOLOGICAL PUMP removes carbon from surface waters by settling of organic and inorganic carbon derived from biological production to the deep sea. This is a link of the ocean carbon cycle to biological processes.

Understanding the natural processes that affect the global carbon cycle is an important requisite for correctly predicting the effects of global warming. For this we need a sound descriptive and quantitative background in all aspects of chemical oceanography and a good understanding of the coupling between chemical oceanography, tectonics, climate, and physical and biological oceanography. As illustrated in the figure below the oceans are in continues contact with the atmosphere, lithosphere and biosphere.

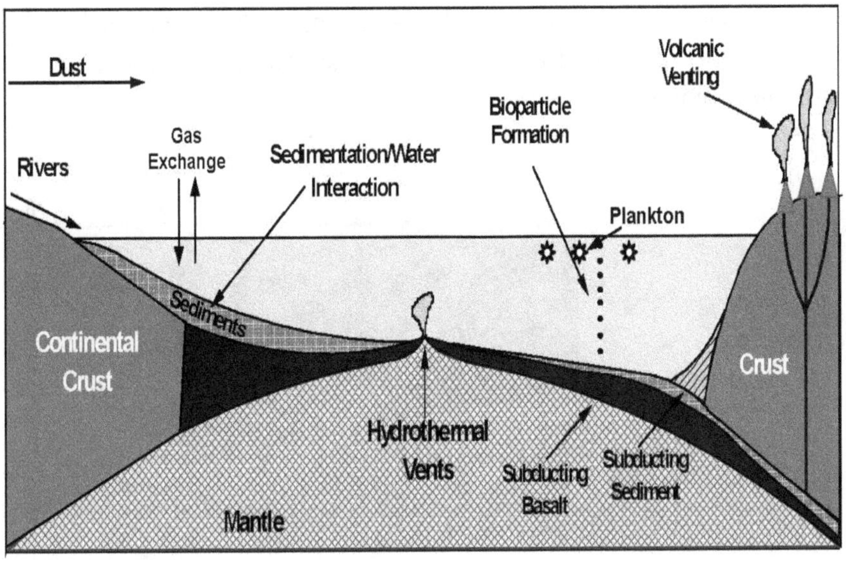

In addition to the major role the ocean plays in the global carbon cycle the world's ocean is also a resource for minerals, energy (gas and petroleum), fisheries, and is the ultimate water source. Chemical Oceanography is the most interdisciplinary of all the sub-disciplines of ocean sciences. The above figure illustrates some of the interactions between the oceans (hydrosphere), lithosphere, atmosphere and biosphere:

- Chemical components of the ocean influence the density of seawater and thus effect its circulation.
- Biological processes in the ocean are controlled by the chemistry (nutrient availability). At the same time biological processes are an important control on chemical distributions. The synergy between biology and chemistry has led to a whole new thriving sub-discipline called biogeochemistry.
- Chemical components are tracers of physical, biological, geological and chemical processes. Understanding what controls chemical distributions helps us understand ocean dynamics.
- Oceanic – Crustal coupling control the distribution of many ions in seawater on time scales of 10^4 to 10^6 years. Thus, we can learn from ocean chemistry about weathering, hydrothermal activity, and other crustal processes.
- Chemical components of marine sediments provide clues necessary to unravel the history of past ocean chemistry and ocean-atmosphere dynamics. Understanding the past should help us predict the future.

Some Descriptive Oceanography

The topography and structure of the ocean floor are highly variable from place to place and reflect tectonic processes within the Earth's interior. These features have varied in the past so that the ocean bottom of today is undoubtedly not like the ocean bottom of 50 million years ago. Even as short as about 5 million years ago Central America did not exist and there was an open seaway between the Atlantic and Pacific. The major topographic systems, common to all oceans, are the continental margins, the ocean-basin floors and the oceanic ridge systems. Tectonic features such as fracture zones, plateaus, trenches and mid-ocean ridges act to subdivide the main oceans into a larger number of smaller basins.

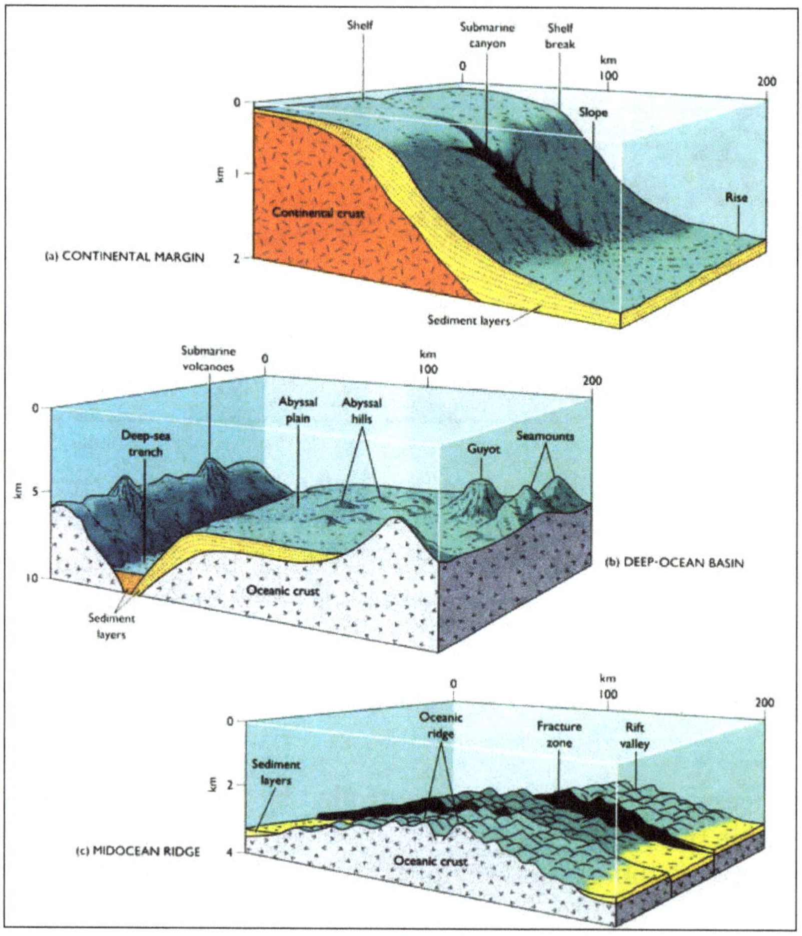

The continental margin regions are the transition zones between the continents and ocean basins. The general features of continental margins may be one of two sequences: shelf-slope-rise-basin or shelf-slope-trench-basin.

The continental shelf is the submerged continuation of the adjacent land, modified in part by marine erosion or sediment deposition. The seaward edge of the continental shelf can frequently be clearly seen and it is called the shelf break. The shelf break tends to occur at a depth of about 200 m over most of the ocean. On average, the continental shelf is about 70 km wide, although it can vary widely (compare the east coast of China with the west coast of Peru).

The continental slope is characterized as the region where the gradient of the topography changes from 1:1000 on the shelf to greater than 1:40. Thus continental slopes are the relatively narrow, steeply inclined submerged edges of the continents.

The ocean trenches are the topographic reflection of the subduction of oceanic plates beneath the continents. The greatest ocean depths occur in such trenches. The deepest is the Challenger Deep, which descends to 11,035 meters in the Marianas Trench.

The continental rises are mainly depositional features that are the result of coalescing of thick wedges of sedimentary deposits carried by turbidity currents down the slope and along the margin by boundary currents. Deposition is caused by the reduction in current speed when it flows out onto the gently sloping rise. Gradually the continental rise grades into the ocean basins and the abyssal plains.

The average depth of the oceans is 3730 m. The area, volume and average depth of the ocean basins and some marginal seas are given below. The Pacific Ocean is the largest and contains more than one-half of the Earth's water.

Region	Area (10^6 km^2)	Mean Depth (m)	Volume (10^6 km^3)
Pacific Ocean	165.25	4282	707.56
Atlantic Ocean	82.44	3926	323.61
Indian Ocean	73.44	3963	291.03
Arctic	14.09	1205	16.98
South East Asia Seas	8.14	1212	9.87
Central America Seas	4.32	2216	9.57
Mediterranean Sea and Black Sea	2.97	1429	4.24
Hudson Bay	1.23	128	0.16
Other Marginal Seas (Baltic, Red etc.)	9.18		7.3

Global Water Cycle

Water is present on Earth in three phases - solid, liquid and gas. The ocean contains the bulk (97%) of the earth's water (1.37×10^9 km^3 or about 1.37×10^{24} g) and moderates the global water cycle. Glaciers are the second largest reservoir with about 2% of the total. The reservoir of water in rivers, lakes and the atmosphere is a trivial part of the total (0.003%). A summary of the water reservoirs is given below:

Environment	Water Volume (km^3)	Percentage of Total (%)
Surface Water	125,000	
Freshwater Lakes	104,000	0.01
Saline Lakes and Inland Seas	1,200	0.009
Rivers and Streams	230,000	0.0001
Total		0.0191
Subsurface Water		
Soil Moisture	67,000	0.005

Ground Water (shallow, <750m)	4,000,000	0.30
Ground Water (deep, 750-4000m)	5,000,000	0.38
Total	9,067,000	0.685
Ice Caps and Glaciers	29,000,000	2.05
Atmosphere	13,000	0.001
Biosphere	600	.00004
Oceans	1.37×10^9	97.25
Total	1.408×10^9	

Water is continually moving between reservoirs as part of the hydrological cycle. These fluxes are summarized in the figure and table below. Evaporation exceeds precipitation over the ocean, while precipitation exceeds evaporation over land. River flow from land to the ocean accounts for the difference.

Flux	Water Flux (km^3 yr^{-1})
Ocean Evaporation	423,000
Ocean Precipitation	385,600
Land Evaporation	72,900
Land Precipitation	110,300
Runoff from Land	37,400

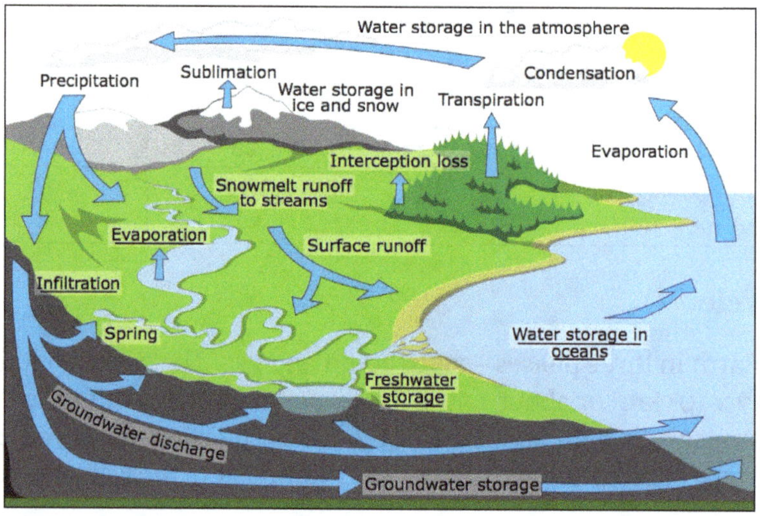

The ocean has a turnover time of about 37,000 years with respect to river inflow. This is how long it would take to fill the ocean if it were totally dry. Turnover times are defined as the mass in the reservoir divided by the input or removal rate. By comparison the average residence time of water in the atmosphere with respect to evaporation from the oceans and continents is only about 10 days.

The ocean's role in controlling the water content of the atmosphere has important implications for past, present and future climates of the Earth. Water vapor itself is the most important greenhouse gas and, alone, is responsible for about 2C of greenhouse warming. Without any greenhouse gases the average earth temperature would be 26K. Instead it averages 28K mostly because of the trapping of infrared radiation by water vapor. Water's unusually high heat capacity and latent heat of evaporation play an important role in heat storage and transport. Here are a few examples of some of the key questions addressed in chemical oceanography.

Why is the ocean salty and what controls its salt content? Does evaporation of river water make seawater? The answer is no. Evaporation of river water makes an alkaline lake with pH values ~10 and very low concentrations of calcium and magnesium:

- What are the different forms of chemicals in the oceans and how can we calculate their concentrations using a systematic chemical approach? Copper is toxic to phytoplankton like Thalassiosira pseudonana. The growth rate of this organism decreases with increasing total concentration of Cu, but with much scatter in the data (left figure below).
- When the speciation of copper is taken into account and the activity of the free Cu^{2+} ion is calculated, the scatter is removed. The conclusion is that the activity of Cu^{2+} and not the concentration is the main determinant of the toxicity.

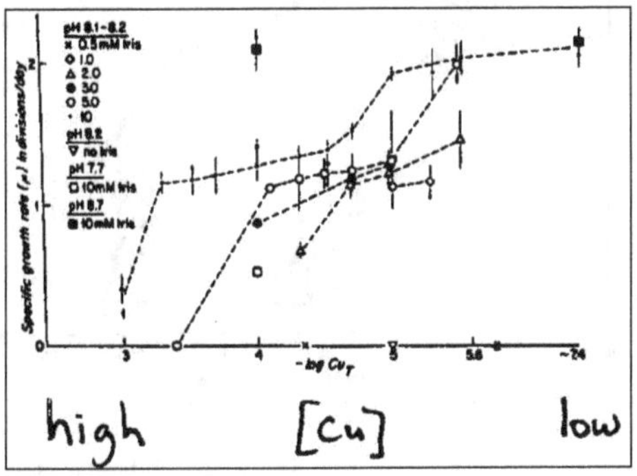

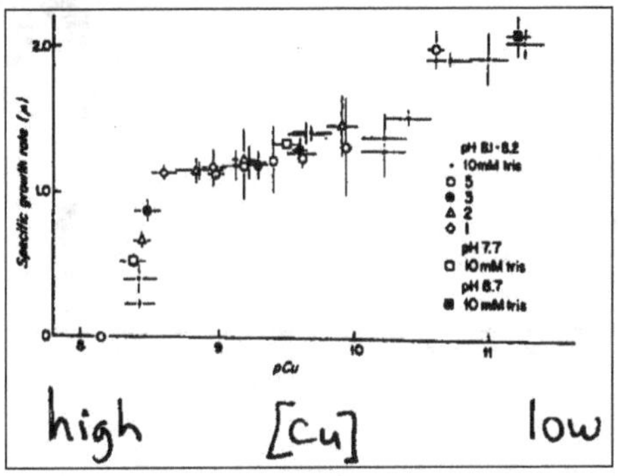

- What are the chemical controls on biological production and food web structure of the ocean? In some ocean areas iron limits phytoplankton growth. Martin conducted experiments where he added iron to samples and showed that adding iron stimulated growth rate and also stimulated diatom production.

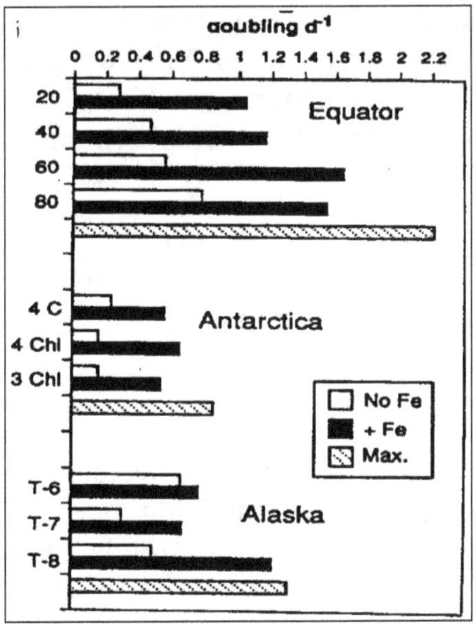

- In what ways do biological processes control the distributions of chemicals in the ocean? Photosynthesis and respiration control the distribution of many elements in seawater.

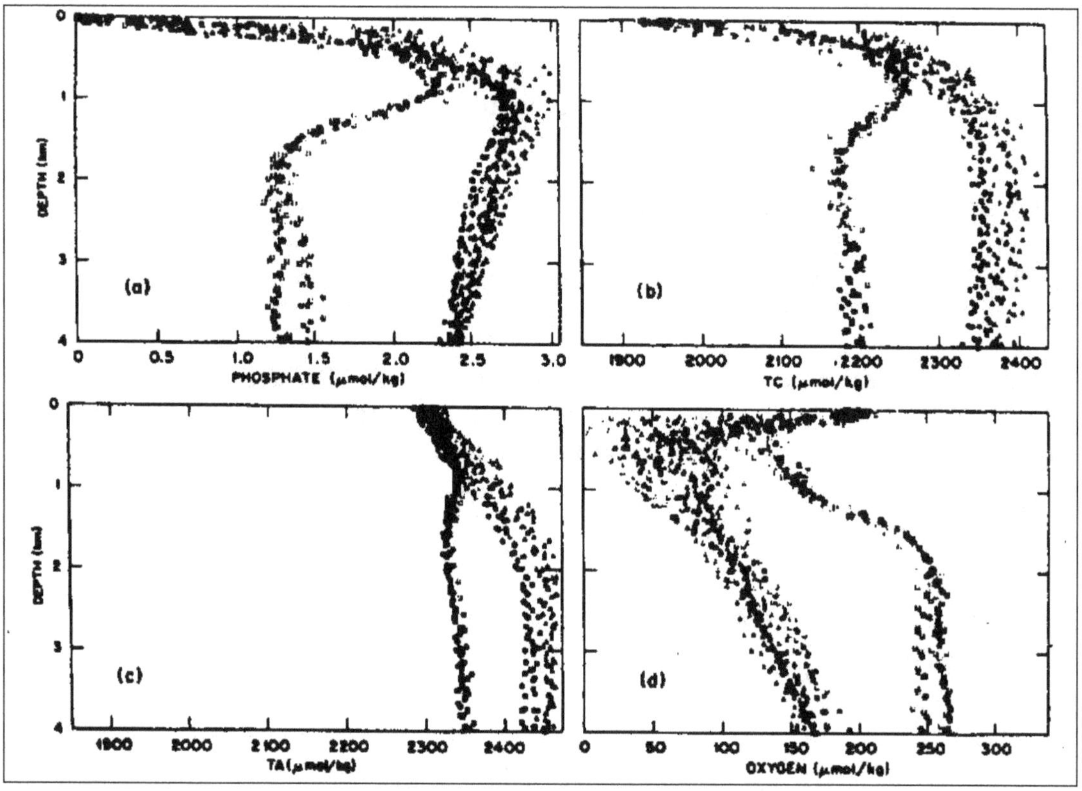

- What controls the distribution of clay minerals and biogenic phases like calcite ($CaCO_3$), opal (SiO_2) and organic carbon in marine sediments? $CaCO_3$ is enriched in shallow sediments and almost totally absent in deep sediments. A balance between the production of $CaCO_3$ by organisms in the upper water column and dissolution in deeper waters, which is a function of temperature and pressure, controls $CaCO_3$ distribution.

- What can chemical tracers (specific elements and compounds as well as stable and radioactive isotopes) tell us about physical, biological and geological processes and their rates in the ocean? For example, the distribution of ^{14}C and CFCs in seawater can tell as when a water mass was last in contact with the atmosphere thus, the "age" of that water mass and circulation rates.

- How important is the ocean as a sink for fossil fuel CO_2 and other greenhouse gases? Brewer calculated the pCO_2 in samples from the core of the salinity minimum of the Antarctic Intermediate Water. The youngest samples from further south have highest pCO_2 reflecting the increase in atmospheric CO_2.

- How much has atmospheric CO^2 varied in the geological past and what were the controls? Martin shows the distribution of iron and CO^2 versus age (0 to 160,000 years BP) in the Vostok ice core in Antarctica. Martin argued that there was increased atmospheric input of dust rich in iron during the last glacial. This stimulated biological production in the ocean and lowered atmospheric CO_2.

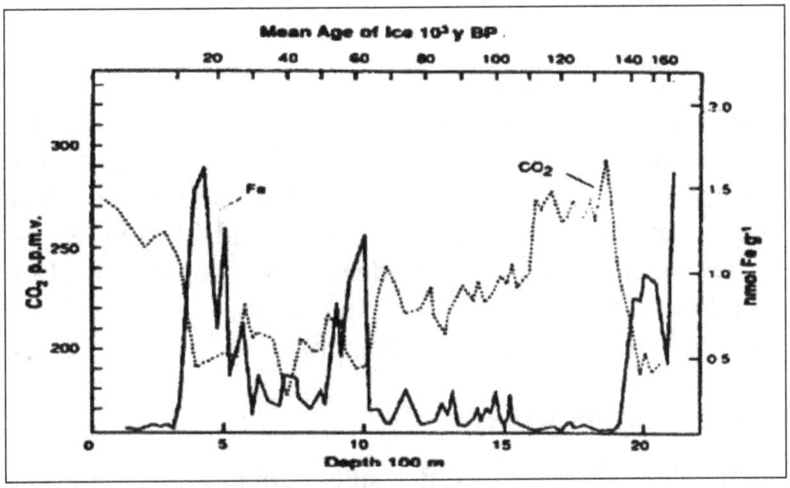

Biological Oceanography

Biological oceanography is a major scientific discipline dealing with all aspects of marine life under different zones of the oceanic environments. The interest to study biology by humans started as early as fourth century BC when Aristotle described about 180 species of marine animals. The geographical knowledge of oceans got improved after several great sea expeditions conducted by the people from 15th to 16th centuries. Through Ocean explorations people conducted detailed underwater surveys and mapped the ocean floors with respect to their physical features, chemistry and biological conditions.

The First Texts of Oceanography

The first texts of Oceanography were published by Charles Wyville Thomson in 1873, entitled as the "The Depths of the sea". Through the Challenger Expeditions of 1872, people traveled for 110900 km and visited all the major oceans of the world, except Arctic. The Challenger expedition also attempted to integrate the geology, biology, chemistry and the physico-chemical phenomena o the oceans. It was carried out for about 19 years by a group of 76 scientists. The first seafloor ma was produced with details of life existing at greater depths. About 715 new genera and 4417 new species of marine organisms were described by the great German Biologist Ernst Haeckel. Since 1872, different countries have been engaged in conducting many major biological oceanography expeditions. The study of marine life has offered more and more data and scientific facts about the dynamics of oceanic environments.

Edward Forbes, a British Naturalist is the founding father of Oceanography. He has systematically studied the marine biota and benthic marine animals. It was he, who specified that different biological species occupy different depth zones of the seas and oceans. His Nephew, James Ross, during 1839-43, collected the samples of benthic animals as deep as 730 m, and gave a lot of information to others about the existence of marine life at deeper levels. When facilities were very limited to carry out any underwater explorations, this was done. Ecology of marine life was understood and it became a major part in the study of oceanography and biology.

Component of Marine Environment

Marine ecosystems are characterized by both biotic and abiotic components. The dominant biotic components are organisms and their species, predators, parasites, competitors and Mates. The dominant abiotic components (i.e. the physical and chemical components) are temperature, concentration of nutrients, penetration of sunlight, turbulence and turbidity, salinity and density of water masses, climatic parameters including the action of wind. Marine ecosystems are typical environments ranging from a small tidal inlet to the deep water ocean basins. Depth of water column is a major factor in addition to its movement and circulation. Solar and atmospheric interactions are very important aspects in the biological oceanography.

Conditions of Marine Environment

The marine environments show altogether a different kind of physical condition for their life to survive. The major factors influencing the oceanic waters are their temperature, transparency, salinity and density which vary with reference to space, time and depth. The ocean water temperature decreases and light penetration diminishes with reference to depth in oceans. The hydrostatic pressure of water increases with depth. The nutrients become more concentrated when the depth increases.

Conditions of Marine Life

Water, which is a fundamental constituent of all living organisms, is available in plenty, within the marine environments. Marine life species are buoyed up by moving water and need not have to store a large amount of energy in their skeletal material. Majority of the marine plants are also floating species. Size-wise, they are microscopic. Many of the marine animals are invertebrates. They do not possess massive skeletons. For the purpose of floating and swimming, the marine animals require very little energy.

Temperature and Light Penetration

Temperature of oceanic waters is an important parameter which does not vary as drastically as seen on the land masses or as seen in the atmosphere. However, there are certain properties which are less favorable for life to survive in the seas and oceans. The growth of plants in the sea is limited by the availability of sun light. It is a fact that 50% of the total solar radiation which penetrates into the sea surface gets disappeared rapidly with reference to depth. Much of the marine environment is under perpetual darkness. Under such circumstances, the entire marine life depends on the availability of essential nutrients only. It also adds to the release of decayed organic matter, in huge volume, within the seas and oceans. Much of the decaying matter sinks inside the seas and gets mixed up or deposited. The zone just below the sea surface experiences the maximum environmental fluctuation. This is the zone where more air-water interactions exist. The same surface zone experiences much variation in temperature, salinity and turbulence of water from the winds. All the environmental parameters also show distinct vertical variations within the oceanic water masses. Because of these variations, it is necessary to classify the marine zones for any detailed analysis.

Classification of Marine Zones

Marine environments can be subdivided into two major divisions as Pelagic and Benthic environments. The word “Pelagic” means ‘Open sea’ and the word “benthic” means ‘bottom’. The Pelagic condition refers to the ocean water column starting from the surface of the oceans and ending at the greatest depths. The Benthic condition refers to the conditions of the deep ocean floors. In the open sea, the zone extending from the high water and low water coastal belt, up to a depth of 200 m inside the sea is called as the Neritic zone.

Pelagic Zone

The Pelagic zone is further subdivided into five major layers, as epipelagic, mesopelagic, bathypelagic, abyssopelagic and hadal zones. The Epipelagic zone is the zone of water column that exists up to 200m down from the ocean surface. The Mesopelagic zone is the zone that exists up to 1000 m down below the epipelagic zone. The Bathypelagic zone is the underwater zone that is existing from 1200 m to 4000 m down below the ocean surface. The Abyssopelagic zone is the zone that is existing beyond 5200 m depth and up to 6000m from the oceanic surfaces. The Hadal Pelagic zone is the zone of deep water extending beyond 6000m and may go up to 10000 m below the ocean surface. Biological oceanography deals with the physic-chemical and biological conditions of marine life of all these zones.

Benthic Zone

The Benthic environments are classified into five zones as supra littoral, littoral, sub-littoral, bathyal, abyssal and hadal types. The Supra littoral zone is the zone existing on the beach with the high tide water line. The Littoral zone is the high water to low water tide line region. The Sub littoral zone refers to the zone of low tide water line to 200m depth on the continental shelf region inside the sea. The Bathyal zone is the zone ranging from 200 m to 3000 m depth of water column inside the sea. The Abyssal zone is ranging from 2000m to 6000m depth inside the sea and it exists mostly on the continental slope regions. The Hadal zone is the life zone existing beyond 6000 m of water depth. This zone may go up to the end of 10000+ m depth of water column in the deep ocean basins. This is the last and dark zone of the oceans. Biological oceanography deals with all the living conditions of marine life and their biological processes prevailing in all these zones.

Ecology of Seas

The Pelagic environment supports the life of all Planktons and Nektons. It is seen that the epipelagic zone is almost a sunlit zone. Enough amount of sun light penetrates into the top layer of oceans. This situation helps all floating plants to carry on their photosynthesis. The mesopelagic zone is known as the twilight zone. This is a dim zone where there is very little light penetration. This condition is not suitable for many plants to grow. The bathypelagic zone is known as the midnight zone. This is the layer between 1000m and 4000m depth below the sea surface, where there is no penetration of light. The abyssal zone is the pitch-black bottom layer of the oceans. The oceanic water masses present in this zone are also at freezing temperature and with great pressure. The Hadal zone is the last deepest zone. This is the most inhospitable zone of the oceans. Biological oceanography deals with all these zones, prevailing marine life and their biological processes.

Life in Oceans

Plants are found only in the sunlit zones where there is sufficient light for photosynthesis. Animals are found at all depths of the oceans though their numbers are greater near the surface where food is available in plenty for them. More than 90% of all the marine species dwell on the ocean bottom where a single rock can be a home to over ten major groups of organisms like corals, mollusks and sponges. Almost all marine life depends directly or indirectly on the microscopic algae that are found only at the surface zones of the seas and oceans. Hence, most of the animals in the oceans live in the sunlit zone. They migrate towards this zone regularly in search of food. Some animals only eat plants. Biological oceanography deals with all the marine life and their migration patterns.

Study of Controlling Factors

The ecology of seas depends on the biotic and abiotic physical and chemical parameters. The nature of sea water, properties of sea water, interaction of sea water with the atmospheric air, the solar radiation and its effects, salinity and density differences in oceanic waters, the wind driven forces and the changing hydrostatic pressure with reference to depth, are the abiotic aspects controlling the ecology of marine environments. The significant roles played by these parameters are studied under biological oceanography.

Solar Radiation

Sunlight is the most important parameter for all marine life in the seas. Only a fraction of sunlight can penetrate thought the surface zone and help in carrying out the photosynthesis by plants. The energy is consumed for the conversion of inorganic matter to organic compounds. Some amount of radiation is absorbed by water molecules and converted to heat. This heat controls the temperature variations of the oceans. The depth-wise distribution of plants and animals is also controlled by the penetration of light. There is also a periodic change, as observed in the solar radiation and light penetration. These help the vision in animals, their migration and breeding periods within the sea. Biological processes enacted by sunlight are major aspect in biological oceanography to understand.

Radiation at the Surface

Solar radiation at the sea surface and the vertical variation in the light intensity are to be measured in proper units. The notable light units used for biological studies of oceans are:

- Einstein (E) unit, which measures the Photons.
- Watt (W), which measures the energy of radiation.

The energy of radiation depends on the wavelength of the light. Photosynthesis radiation happens between 400 mm and 700 mm. An Einstein is a unit defined as the energy in one mole (6.022×10^{23}) of photons. Because energy is inversely proportional to wavelength, the unit is frequency dependent.

Temporal Variations

About half of the solar energy is absorbed (or) scattered by various layers of the atmosphere. Only 50% of it is reaching the sea surface. In this process, some portion is reflected back to the atmosphere depending upon the angle of incidence. The amount of radiant energy reaching the surface is a function of the sun's angle of incidence, the length of the day, time of the day, time of the year, the latitude, and the prevailing weather conditions. This is called as temporal variations. The temporal variations of radiation may be:

- Diel variation: Change that happens between a day and night.
- Diurnal variation: Change that may occur only during the day time.
- Seasonal variation: Change that happens between the seasons, especially at high altitudes.

Penetration of Light

Based on the light intensity and relative penetration of light in the sea, three vertical ecological zones have been identified as:

- Euphotic zone.
- Disphotic zone.
- Aphotic zone.

The shallowest zone is the euphotic zone existing just below the sea surface. In this zone, light penetration fully supports all phytoplanktons to grow and reproduce. The respiration loss is also balanced by a compensation depth which marks the lower boundary of Euphotic zone. The dim lighted zone below the euphotic layer is called as the disphotic zone. The last layer which is under complete darkness, where no sunlight can reach, is called as the aphotic zone. Biological oceanography attempts to study all the conditions of these zones.

Factors Controlling Circulation

Many marine processes are controlled by the temperature of water. They may be physical, chemical and biological processes. Temperature and salinity of oceanic waters determine the density. Due to this, the vertical water circulation and movements are fully controlled by all these three properties. Exchange of heat happens between ocean and atmosphere continuously. There is also a wide range of temperature variation on the sea surface. It exceeds 30 °C in tropical oceans, 40 °C in shallow seas, and as low as – 1.9 °C in Polar Regions. Oceans are cooled by evaporation. A good amount of heat is transferred during this transformation of water into water vapor.

Classification of Zones with Temperature Variations

Ocean surface temperature fluctuates with reference to days, months, seasons and years. It also varies with reference to Polar, tropical, subtropical and temperate zones of the globe. The surface turbulent waters also transfer heat downwards. Due to this, the uppermost part of Oceanic water has a relatively raised temperature. This gets decreased at a depth of 200 to 300m and up to 1000m.

The water layer showing the steepest temperature gradient is known as thermocline. The zone showing the rapid changes in density of water is known as Pycnocline. Pycnocline acts like a barrier to vertical water circulations, and also animal movement. At 2000 – 3000m depth, the oceanic water temperature never rises above 4 °C. It also goes down up to 0 °C to 3 °C in deeper zones.

Biological Oceanography becomes an interesting subject when we start studying these deep water horizons.

Seawater is Unique

Seawater is unique water. It contains more dissolved salts than river water, lake water and rain water. The salinity is expressed in terms of total dissolved inorganic ions and other compounds and gases. The average salinity of the ocean water is 35 ppt. When surface water gets evaporated, the salinity is increased. It happens at the ocean surface. The Salinity gets lowered due to rainfall, river water inflow and after every snow melts, in the waters of seas and oceans.

Vertical Variation in Salinity

Vertical variation of salinity is an important limiting factor for the marine life to survive. The layer at which a rapid change in salinity occurs is known as Halocline. Salinity varies with reference to seasons, depth and locations. Marine life has unique physiological mechanisms to cope up with these salinity variations. Osmoregulation is one of the essential mechanisms. The marine life is also classified based on their tolerance level of salinity. The species which can tolerate a wide range of salinity are called Euryhaline and those which can tolerate a narrow range of salinity are called as Stenohaline species.

Density of Ocean Waters

The density of seawater depends mainly on the salinity and temperature and to some extent by its hydrostatic pressure. When salinity increases, density also increases. The movement of water masses, in the oceans, is controlled by all these 3 parameters. Horizontal water movement is controlled by wind, temperature and Salinity. The Vertical movement of water is controlled by temperature, salinity and density. The normal density of Seawater at the surface level is very low. When it increases, the water mass sinks down below and reaches the appropriate strata of matching density. The sinking of water is called as down welling and upward movement of water is called as upwelling. Advection is the term used for horizontal and vertical movements of water in these water bodies. Biological oceanography attempts to analyse all these aspects as well.

Pressure in Ocean Water

The hydrostatic pressure of oceanic water column is determined by the equation Sgh, where g is the acceleration due to gravity, S is the density of water and H is the thickness of water column. Pressure increases due to weight of overlying mass. It is expressed in Newtons per Sq. metre. At 10m depth, the pressure will be ten to the power of five Newtons per meter square (10^5 Nm^{-2}). This is equal to 1 bar (or) 1 atmosphere. Hence, the depth wise variation in pressure is calculated based on this range.

All marine life existing within the deep oceans are subjected to very high pressures. The pressures may go up to 1000 atmosphere in the deep ocean basins. Some animals may travel up and down for several hundred metres and experience this change. Both pelagic and benthic species do inhabit these pressure changes. They are called as eurybathic species. The marine life which cannot tolerate such pressure variations are called as stenobathic species. Study of Biological oceanography becomes more interesting while dealing with all these aspects.

Surface Water Currents

Ocean water surface currents are generated by the winds. These water movements influence the biological productivity and nutrient availability. Due to this, the geographical distribution patterns of pelagic and benthic marine species are also varied. Ocean surface currents are fully controlled by global wind systems. Their directions are modified by the earth's rotation. Biological oceanography considers all these aspects while analyzing the marine life.

Marine Flora

Phytoplanktons and Zooplanktons are the two major types of planktons existing in the shallow depth water layers of the oceans. Most of the phytoplanktons are unicellular algae. They also include diatoms, green algae, yellow-green algae, blue green algae, red algae, silicoflagellates and dinoflagellates. In addition, Cryptomonads, Prasinomonads, Chloromonads and Chrysomonads are the other marine phytoplanktons predominantly seen in the seas and oceans. These are present throughout the lighted layers of the oceans. Phytoplaktons are the major primary producers in the pelagic zone. The rate of primary production in plant mass is called as Primary Productivity. The total weight of all organisms in a given area (or) volume is known as the biomass.

Controls of Primary Productivity

The Physical controls of Primary Production are very essential factors. The properties that are controlling the primary production in oceanic shallow waters are Light, Physical forces, Abundance of nutrients and temperature. It also varies with reference to seasons, depth and location.

Marine Fauna

Marine animals are divided into 3 groups as Zooplankton, Nekton and Benthos. Zooplanktons are the drifting animals and are usually small but grow to fairly large size. Typical example is a Jellyfish. The zooplankton population includes some members like fish eggs or larval forms of organisms which may grow up and leave the planktonic community to join the nekton or benthos.

Pelagic Organisms

Pelagic organisms are classified into various types based:

Virio Plankton	0.02-0.2mm – Femto plankton
Bacterio Plankton	0.2-2.0 mm – Pico plankton
Myco plankton	2.0-20.0 mm – Nano plankton
Phyto plankton	20-200 mm – Micro plankton
Protozoo plankton	20-200 mm – Micro plankton
Metazoo plankton	0.2-20 mm Meso plankton
	2.0 to 20 cm Macro plankton
	20-200 cm – Mega plankton

A bucket of seawater might hold a million microscopic diatoms which are relatives of sea weeds encased in glassy boxes. To grow, phytoplankton needs nutrients from the seawater and lots of sunlight. The large quantities of diatoms and phytoplankton give a color to the sea water.

General Types of Plants

There are two general types of plants found in the ocean. They are those having roots that are attached to the ocean bottom and those that are not having roots which simply drift about with water. The rooted plants are only found in shallow water because of the availability of sunlight for photosynthesis. The most abundant plants in the ocean are the phyto planktons. These are usually single-celled, minute floating plants that drift throughout the surface of the oceans.

The word Plankton means wandering or drifting in Greek. Planktons are floating plants or organisms. There may be Phytoplanktons denoting plants and Zooplankton denoting the animals. Planktons are mostly microscopic dimension. Nektons are free-swimming animals. Some of the Phytoplanktons are passively transported by the currents in the sea. Fish, squid and marine mammals are the major Nektons, of oceans.

Nektons are the free swimmers and the largest portion of familiar population of animals found in the oceans. Common fishes, octopus, whales, eels and squid are all examples of nekton. Whales, Sea mammals, dolphin and porpoise codfish/trout. The third type of sea animals spends their entire life on or in the ocean bottom. This group of marine animals is called benthos. Lobsters, starfish, various norms, snails, oysters, etc.

All marine life needs food. Plants make their own food but animals obtain food from their environment. A food chain represents the transfer of body-building substances and energy when one organism eats another. Diatoms form the first link in the marine food chain. In the ocean, there are innumerable individual food chains overlapping and intersecting to form complex food webs. Most marine creatures eat a variety of foods. The rich diversity of life in the sea forms a delicately balanced. Network of predators and prey. All organisms are dependent on one another for survival.

Marine Zooplanktons

Marine Zooplanktons animals range from microscopic unicellular organisms to jellyfish size which is several metres in diameter. They are of two kinds as:

- Holoplankton (Permanent Planktons-Spend their entire life cycle in water column).
- Meroplankton (Temporary residents of Plankton community).

Planktons show vertical migration within every 24 hours. This is called as diel vertical migration. This occurs in many epipelagic and mesopelagic species.

Holoplanktons

There are about 5000 species of holoplanktons existing in seawaters. Many notable ones are Foraminifers, Radiolarians, Ciliates, Tintinnids, Jellyfish, Cnidarians, Ctenophores, arrow worms, and some heteropods, Pteropods Copepods, Ostrocods, Amphipods, Euphansiids, Mysids, Decapods and Salps.

Meroplanktons

The Meroplanktons are miniature adults spending a few minutes to several months and years in the upper layers. Most of them are larvae of benthic forms. They include Snail religers, starfish, sea urchins, barnacles and crabs.

Migration of Marine Animals

Marine Zooplanktons show three kinds of patterns as:

- Nocturnal migration - After sunset.
- Twilight migration - Rising up at sunrise.
- Reverse migration - Surface rise during the day and descent to depths at nights.

There are also seasonal vertical migrations happening in the shallow waters for marine life. Zoogeographic studies help understanding the distributions of living organisms and the physiological or ecological reasons behind these mobilities. Long –term observations show that plankton abundance and species composition may change with reference to time, climatic variations, stratification and nutrient availability.

Properties affecting marine life

Properties affecting the marine life are many. The following are the properties affecting life in the seas and oceans:

- Quality of sea water.
- Skeleton shape and size of organisms.
- Buoyancy.
- Gravity.
- Temperature of ocean water.
- Density.
- Light.
- Availability of Nutrients.
- Water turbulence.
- Hydrostatic pressure.

Aspects of Biological Oceanography

In addition to these, biological oceanography deals with the study of:

- Migration.
- Zoogeography.
- Energy Flow and Mineral Cycles.
- Mineral cycles.

- Marine mammals/ Sea birds / Marine Fish-Fish migrations.
- Fisheries –world Fish catch/ Mari culture.
- Coral Reefs, Human Impacts on Marine Biotic.
- Marine Pollutants:
    - Petroleum Hydrocarbon.
    - Plastics.
    - Pesticides.
    - Heavy metals.
    - Sewage.
    - Radioactive wastes.
    - Thermal effluents.

Coastal Oceanography

The coastal ocean links the continents and the open (deep) ocean; as such, it overlies the continental margin that consists of the shallow continental shelf, steep continental slope, and the deep continental rise. These strong variations in water depth, and associated geomorphological anomalies (for example, submarine canyons and capes and submarine banks and basins), together with earth rotation, exert controls on the direction of coastal ocean circulation. Density stratification, friction, and transience mediate these constraints. Thus, coastal oceanic processes account for the ultimate cross-shore transport of materials from the continents to the open ocean, but which is often delayed and extended through circuitous alongshore displacements influenced by topographic controls.

The coastal ocean is very responsive to atmospheric forcing associated with the "weather cycle" and serves as a coastal wave guide providing teleconnections that propagate disturbances hundreds or even thousands of kilometers alongshore. It is also under the influence of forcing from the tides, river runoff, meandering oceanic jets, and mesoscale eddies. Consequently, the coastal ocean is highly variable and, not coincidentally, ecologically productive, supporting most of the major fisheries of the world.

Fortuitously, the scientific understanding, and supporting technologies, needed to address coastal ocean processes and regional-scale analyses are emerging just as the societal demand has raised for a higher level of coastal environmental and ecological information services to meet the need for integrated coastal zone management (ICZM) and numerous other applications.

"Coastal oceanography" embraces the study of the physics, chemistry, ecology, and geology of the "coastal ocean." The "coastal ocean" basically encompasses the 200 nautical miles (NM) wide

Exclusive Economic Zone (EEZ), a political concept introduced through the United Nations Law of the Sea Treaty (UNLOS) that has practical consequences for ocean resource management, both for exploitation and conservation. It also has consequences for the sponsorship of basic and applied research, and the development and implementation of operational observing and modeling systems. Furthermore, utilization of the EEZ as a defining principle for the coastal ocean is useful scientifically because it more or less coincides (not by chance) with the continental margin; that is, the continental shelf, continental slope, and continental rise that serve to structure the dynamics of the coastal ocean. Usually the estuaries, coastal lagoons, Great Lakes, and inland seas are also included in the definition of the coastal ocean. The marginal and semi-enclosed seas are special cases of the coastal ocean, and are thus treated as well. Here, the topical coverage is focused on the physics of the coastal ocean, with some references to implications for ecology, chemistry, and geology where most appropriate.

General Nature

The coastal ocean is where the continents meet the open ocean; hence, it serves as a "bridge" for transporting organic and inorganic, natural and anthropogenic materials between the land and sea. More accurately, it functions as a turbulent boundary layer that joins the land and the sea under the influence of forces and their responses (involving a variety of phenomena and processes) on a large range of time and space scales. As a beginning example, the coastal ocean is impacted by El Niños (with time scales of several months and space scales of a few mega meters), as well as surface gravity waves (with time scales of several seconds and space scales of a few hundred meters). Similarly, the coastal ocean circulation is driven by a variety of (often competing) forces: winds, atmospheric pressure, solar and atmospheric heating/cooling, evaporation/precipitation, freezing/melting, buoyant river discharge, tides, and offshore currents and their meandering jets with associated mesoscale eddies and fronts. Some of the forcing is hourly; other forcing is seasonal or even interannual, interdecadal, or longer.

The response of the coastal ocean to the various forcing functions depends upon the joint effects of the earth's rotation (that is, the Coriolis parameter), variable bottom topography, and density stratification, and it is manifested in various processes and phenomena. In addition to the forced responses noted above, the circulation of the coastal ocean can be controlled by its intrinsic variability because of dynamical instabilities of the currents. Hence, the coastal ocean is generally characterized as being highly variable in space and time. Indeed, in some circles several decades ago, the coastal ocean was thought to be too complicated (due to its intense variability over a broad range of time and space scales) to be amenable for study. As a result of modern instrumentation, theoretical and observational advances in process understanding, and developments in numerical simulation, a degree of order has been brought to the investigation of coastal ocean regimes that had been previously thought perhaps overwhelmingly chaotic, and that had been approached with hopelessly naive methods, deficiently idealized concepts, and grossly under-sampled observations.

Importance

Historically, knowledge of the coastal ocean has been essential to the safe and efficient conduct of maritime trade, and for the defense of coastal nations and attacks by maritime powers. Interest in tides, storm surges, tsunamis, and long-term coastal sea level variations has been motivated over recent centuries by concerns for coastal flooding, safe navigation, and safety of coastal populations and infrastructure. Coastal marine ecosystems are rich and diverse, supporting much of the commercial and recreational fisheries of the world. Rich deposits of oil and gas, other minerals, and sand and gravel underlie many portions of the coastal ocean. In recent decades, coastal tourism has become a significant component of the world economy, but it depends, in part, on the quality of the coastal environment and the abundance of coastal fish. By their very nature, most rivers discharge to the coastal ocean; their discharges include suspended and bed load sediments. With the rise of modern agriculture, their discharges also include fertilizers and pesticides. With the rise of industrial and municipal waste disposal, their discharges include bacteria and viruses; and with the rise of industrialization, their discharges include various hydrocarbons, toxins, and carcinogens. Some of these discharged chemicals accumulate within the coastal ocean and its ecosystems and sediments; others are transferred to the open ocean, sediments on the deep seafloor, or the atmosphere. With the already high and rapidly growing human population densities in the coastal zones of the world, the coastal nations are under increased pressure to balance the conflicting uses of the coastal ocean. Furthermore, the coastal ocean is host to several natural hazards that threaten human coastal populations, both their infrastructure and human health; for example, storm surges and coastal flooding, storm-driven sea and swell, tsunamis, land-falling hurricanes, coastal erosion and deposition, rising sea level due to climate variability and global change, eutrophication and hypoxia (and its end state anoxia), and harmful algal blooms. Only the more developed coastal nations have a modicum of understanding of their coastal resources and risks, let alone a scientific understanding of coastal oceanography. There is also a great shortage of generally available, continuing, real-time, and strategic observations, and supporting modeling activities, for the coastal ocean. Thus, there is a severe lack of information pertaining to the day-to-day state of the coastal ocean. Consequently, awareness and documentation of the onset and evolution of natural and anthropogenic events are seriously limited, restricting the efficacy of emergency and environmental managers. Logically, it would therefore seem that there is an urgent global priority to address coastal oceanography in its broadest sense.

In this regard, it is becoming increasingly difficult to place bounds on coastal oceanography. For example, the coastal ocean is so strongly forced dynamically, and even chemically, by the coastal atmosphere that knowledge of coastal meteorology needs to be taken into account. Conversely, the coastal ocean provides thermal, and also chemical, forcing (for instance, out-gassing of dimethylsulfide (DMS) formed by certain plankton that plays a role in marine stratus formation and ejection of salt particles that help form marine aerosols) to the coastal atmosphere. Thus, the coupling of the coastal ocean and atmosphere is increasingly considered, especially as coastal oceanic and atmospheric observing and modeling systems adopt mesoscale resolution. As another example, the coastal ocean is forced dynamically and chemically by surface water runoff from rivers, estuaries, coastal lagoons and so-called "non-point" sources along the coastline, and by groundwater flowing into shelf sedimentary layers and then entering the ocean through seeps or springs. Conversely, seawater invades coastal aquifers and spoils them with its brine, which is an increasing problem as coastal populations draw down the water table, thus facilitating such invasions. The linking of coastal oceanography and coastal hydrology is in its infancy, and it cannot be done well without coastal meteorology. However, it can be anticipated that such linking will occur as coupled coastal oceanic, hydrological, and atmospheric processes and modeling systems are addressed, both scientifically and societally. These initial perspectives provide the basis on which this article is built, and their salient points are revisited at the end.

Some Dynamical Factors

Variable Bottom Topography and Its Effects

The coastal ocean overlies the continental margin that consists of the continental shelf, continental slope, and continental rise. The continental shelf extends from the coastline to the shelf edge or shelfbreak, typically at a depth of 100 to 200 m. The width of the shelf ranges from about 10 km to a few 100 km; thus, the bottom slope of the shelf is typically between 0.001 and 0.01. The continental slope has a typical bottom slope of 0.1 and extends to a depth of a few km, where the continental rise continues to greater depths with a gentle slope of about 0.001. With an idealized smooth, monotonically shoaling continental margin, surface gravity waves propagating from offshore to the coast will begin to "feel" (that is, interact with) the bottom at some depth depending upon the incident wave frequency, and the waves slow (and amplify) because their phase speed equals the square root of the product of gravitational acceleration and water depth. As a consequence, if the waves are obliquely incident, they will refract to align their wave crests with the coastline, where they break (that is, form surf) and generate mean alongshore jets near shore.

While most surface gravity waves are generated by winds (in the near field, they are called "wind waves," and in the far field, they are called "swell"), tsunamis are generated by tectonic disturbances on the sea floor, undersea volcanoes, and catastrophic mudslides over the continental slope. These relationships underlie many aspects of coastal ocean dynamics, including the trapping of certain classes of surface gravity waves on the shelf, called "edge waves," that are constrained to alongshore propagation. Here, the scope is not limited to the idealized continental margin and the time scale of seconds to minute's characteristic of surface gravity waves.

The coastal ocean commonly has bottom topographic perturbations that can be categorized as submarine banks or basins, coastal islands or lagoons, submarine canyons or capes, and smaller scale bottom roughness elements associated with sand waves and ripples, rock outcrops, and other geological anomalies. Similarly, coastlines are highly fractal as they are characterized by a series of capes and embayments on all scales; in some regions, they are punctuated by fjords and, in others, they are connected to estuaries or coastal lagoons. Because of its relatively shallow water depth and commonly occurring strong bottom slopes and complex coastline configurations, topographic effects play a prominent role in determining the circulation of the coastal ocean.

For example, the combined effects of variable depth and the earth's rotation tend to constrain the mean circulation to flow along isobaths, that is, as alongshore flow. This constraint also leads to the general expectation of mean cyclonic circulation around basins and mean anticyclonic circulation around banks. However, the effects of density stratification, wind forcing, bottom friction, and time-dependence can break that constraint. Obviously, the constraint of alongshore flow must be broken, at least episodically in time and/or space, to accommodate the cross-shore flux of materials from the continents to the open ocean that is known to occur, but seldom observed.

As another example, in regions where the tides have relatively large amplitudes, the tidal currents are strong, producing strong turbulent mixing at the bottom and associated bottom friction, especially over shoal areas such as banks. More generally, the bottom roughness elements serve to produce turbulence due to currents and wave motions that in turn damp the flow and waves. Actually, there is often a two-way interaction, depending on sediment grain size and composition, with the currents and waves serving to shape the sedimentary bed forms through erosion, re-suspension, advection, and deposition, thus creating or modifying some of the bottom roughness elements.

The irregularities in coastlines cause important effects; for example, an alongshore wind stress typically sets up an alongshore pressure gradient that can only be "supported" in the presence of capes; when the wind stress relaxes, the pressure gradient produces a counter flow as it relaxes. Part of this scenario is the generation of cross-shore "squirts and jets," that may become connected to the filaments (mentioned below) driven by mesoscale eddies over the continental slope, and together they can transport shelf waters a hundred kilometers or more offshore over the course of several days.

While contemporary studies of coastal ocean circulation generally treat the bottom topography and coastline as time-invariant, the fact that the near shore bottom topography and coastlines vary with the passage of storms on a seasonal basis implies that two-way interactions between currents and sediments will need to be incorporated in future studies, especially in numerical coastal ocean circulation models.

Geological Oceanography

Geological Oceanography is a division of oceanography. It is mainly dealing with the basic concepts of lithosphere & hydrosphere. It includes the study of the oceanic crust, continental margins, ocean bottom relief, ocean basins, oceanic ridges, rift-valleys, Island arcs, seamounts, marine sedimentation, geology of corals, beach forms and processes, water masses , factors affecting ocean circulation, waves and currents, tides and energy coastal erosion and drifting of sediments, sea level changes, depositional environments and marine deposits. Geological oceanography is also concerned with the occurrence of oil-traps and energy sources, tectonic movements- underwater eruptions, mud volcanoes and impacts of tsunamis.

The Crust and Plate Tectonics

The crust of the earth consists of rigid moving plates. These plates can move apart and create new crusts. They may slide past each other and create strike slip faults. They may converge on each other by subduction, where a crust is consumed and the area will be shrinking. Most of these movements occur along narrow zones between plates where the results of plate-tectonic forces are most evident. There are four types of plate boundaries. The first one is the divergent plate boundaries. In this kind of boundaries, a new crust is generated as the plates pull away from each other. Perhaps the best known of the divergent boundaries is the Mid-Atlantic Ridge. The next one is the convergent boundaries. In this, the crust is destroyed as one plate dives under another. The next type is the transform boundaries. In this type, the crust is neither produced nor destroyed as the plates slide horizontally past each other. The plate boundary zones are broad belts in which the plate boundaries are not well defined and the effects of plate interaction are also not clear. All these aspects are studied in geological oceanography.

Instrumentation

Instrumentation is vital to the study of physical oceanography. Determining wave height, water temperature, or current flow is impossible without a means of measurement. The height of the waves pounding onto the shore can be measured using conventional type of measuring instruments. Away from the shore, the wave height, in deep water, is measured using an instrument called as tide gage. The tide gage is immersed in the water and measures the weight of the water on top. Measurements in physical oceanography are done over a large range of scales. The measurements of ocean currents can vary from a few centimeters to the entire globe.

Geodetic Measurements

The present day plate movements can be tracked directly by means of ground-based or space-based geodetic measurements. Geodesy is the science of analysing the size and shape of the Earth. Ground-based measurements are taken with conventional but very precise ground-surveying techniques, using laser electronic instruments. Since the plate motions are on a global scale, they are best measured by using the modern satellite-based methods.

Space Geodesy

The late 1970s witnessed the rapid growth of space geodesy. It is a term applied to space-based techniques for making very precise, repeated measurements of carefully chosen points on the Earth's surface separated by hundreds to thousands of kilometers. The three most commonly used space-geodetic techniques include very long baseline interferometry (VLBI), satellite laser ranging (SLR), and the Global Positioning System (GPS). Most of these are based on the technologies that were developed for military and aerospace research purposes. Among the three techniques, the GPS has been the most useful instrument for studying the Earth's crustal movements, today.

Morphology and Relief of Ocean Floors

On the basis of bathymetric and other studies, the depth zones of oceans are divided into two major divisions called as Continental Margins and Deep Ocean Basins. The Continental margins include two major zones as Continental shelf and Continental slope. The Continental shelves are adjacent to the beach zones. It is followed by the Continental slope. The Continental slopes are the inclinations towards the sea. From there onwards, there is a steep descent towards the interior parts of the oceans. The Deep Ocean Basins are the darkest depressions of the sea floor.

Marine Sedimentation

The major marine sediment types of ocean are of four kinds. The Lithogenous or terrigenous sediments are the detrital sediment derived from the continental land masses. The biogenous sediments are the remains of dead organisms. The hydrogenous sediments are derived from the precipitation of dissolved minerals in water. The cosmogenous sediments are the dust and meteorite fragments derived from the Outer Space. The marine sediments of the oceans are sorted by their grain size. The gravels stay close to their source areas like river mouths. The sands move along the coasts by wave action. The silts and clays are suspended in the water column by wave action and are carried out in to the deep sea. All the biota may injest into the fine sediments and makes fecal pellets. Sedimentation patterns are analysed in geological oceanography.

Marine Mineral Resources

Oceans & Seas are the sources for food, energy and mineral deposits. Huge volumes of Mineral matter are deposited in ocean bottoms. It is estimated that 40 pounds of gold and 200 pounds of lead are present in dissolved form in every one cubic mile of ocean water. Vast resources of petroleum, coal and natural gas lie buried deep below oceans. The oceans are the treasure houses for Polymetallic Nodules, Cobalt-Rich Crusts and Polymetallic Massive Sulphides. Minerals derived by mechanical erosion from continental rocks are concentrated as placer deposits. These minerals contain heavy metallic elements like barium, chromium, gold, iron, rare earth elements, tin, thorium, tungsten, zirconium and non-metals including diamonds, lime, siliceous sand and gravel.

Coastal Zone Processes

Coastal zones are very sensitive zones. Continental shelves and near-shore waters comprise about 5% of the area of the world. In spite of this, two-thirds of the world's population thrives near the coasts. Throughout human history, people have been attracted towards the coasts because of the moderate climates, food, commercial and recreational opportunities they provide nearer to the beach. Coastal ecosystems are good habitats for innumerable flora and fauna. Coasts are an interface between the lithosphere and hydrosphere. Time immemorial, beaches are considered to be the best places for tourism, fishing, Mari culture, water transport, recreation, sporting, social, political and community interactions.

Geomorphology of Coasts

Every continent or island is bordered by a long or short coastline. Coastline is the line separating the land and sea. Coastal zones are the transition zones between terrestrial and marine habitat. They form an interface between land and oceanic natural processes. Coastal areas also are varied in their topography, climate and vegetation. Some are sandy beaches, rocky shores, with or without tidal inlets. The climate of a coast are controlled by the land and sea breezes and the humidity controlled by marine water. Waves, tides and currents are very powerful geomorphic agents. The erosional and depositional work of the sea waves can create many spectacular landscapes along the borders of the continents. Studying the coastal landforms are interesting aspects in coastal geomorphology.

Chapter 5

Sedimentology: Structure and Process

Sedimentology is the scientific discipline that is concerned with the study of sedimentary rocks, their chemical and physical properties, formation, transportation and deposition. Sedimentary rocks record the events and processes that shaped the Earth's surface and their study is critical to understand the geological history of the Earth. This chapter deals with this discipline in a comprehensive manner.

Sedimentology is the science that deals with scientific study of sediments. The term 'Sedimentology' was first coined by H. A. Wadell in 1932. Sediment word is derived from Latin 'sedimentum' means 'settling'. Sedimentology is the study of the processes of formation, transport and deposition of material that accumulates as sediment in continental and marine environments and finally converts to sedimentary rocks. It also covers the classification, origin and interpretation of sediments in various dimensions.

What is Sediment?

Sediments are fragments of pre-existing rocks viz. igneous, metamorphic and sedimentary rocks, as well as, form by chemical precipitation also. Sediments which are weathered, eroded and transported are known as 'extrabasinal' while others formed in situ or by chemical precipitation are called 'intrabasinal'.

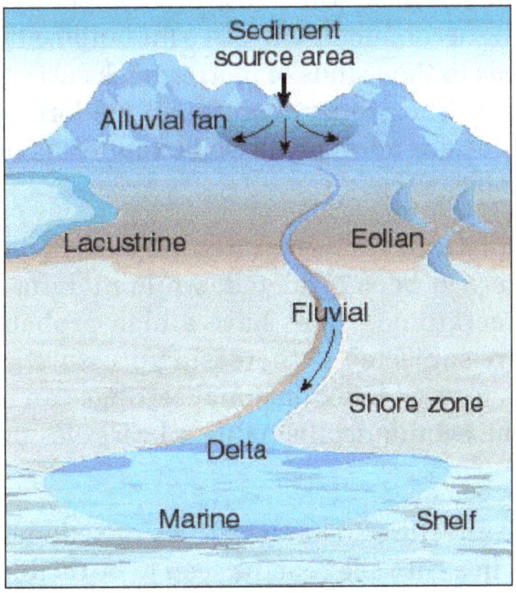

Transport of sediments from sediment source area to different depositional regimes.

Sedimentology is one of the core disciplines of the geology. In the present module, different components of Sedimentology are narrated in brief. Details of these various processes and products are described in separate relevant modules. Different important aspects in the study of Sedimentology are:

- Sedimentary processes.
- Sources of sediments.
- Transport of sediments.
- Classification of sediments and sedimentary rocks.
- Texture and composition of sediments and sedimentary rocks.
- Depositional environment and facies.
- Sedimentary structures.
- Field Sedimentology.
- Petrography.
- Lithification and diagenesis.
- Lithostratigraphy.
- Reservoir properties.

In general term 'basin is a depression, usually of considerable size, which may be erosional or structural in origin'. Sedimentary basins, store house of sedimentary rocks, are regions of the earth of longterm subsidence creating accommodation space for infilling by sediments. Sediments in the basin accumulate into hundreds to thousands of metres thick successions over areas of thousands to millions of square kilometres. The primary control on the formation of sedimentary basins is plate tectonics and hence basins are normally classified in terms of their position in relation to plate tectonic setting and tectonic processes.

Basins formed in different plate tectonic regimes vary in their preservation potential. On oceanic crust, basins are likely to be sub-ducted, while marginal continental basins may be partially preserved, and intracratonic basins have a high probability of preservation. As the sediments are buried, they are subjected to increasing pressure and begin the process of lithification. A number of basins formed in extensional settings can undergo inversion which has accounted for a number of the economically viable oil reserves on earth which were formerly basins.

Each basin type has distinctive features, characteristics of sedimentation and the stratigraphic succession. Like one develops in a rift valley can be seen to be distinctly different from those of an ocean trench. A stratigraphic succession can therefore be interpreted in terms of plate tectonics and places the study of sedimentary rocks into a larger context. The sedimentary rocks in a basin provide a record of the tectonic history of the area. They also provide the record of the effects of other controls on deposition, such as climate, base level and sediment supply. Subsidence in a sedimentary basin results from the thinning of underlying crust, sedimentary, volcanic and tectonic loading, and changes in the thickness or density of adjacent lithosphere.

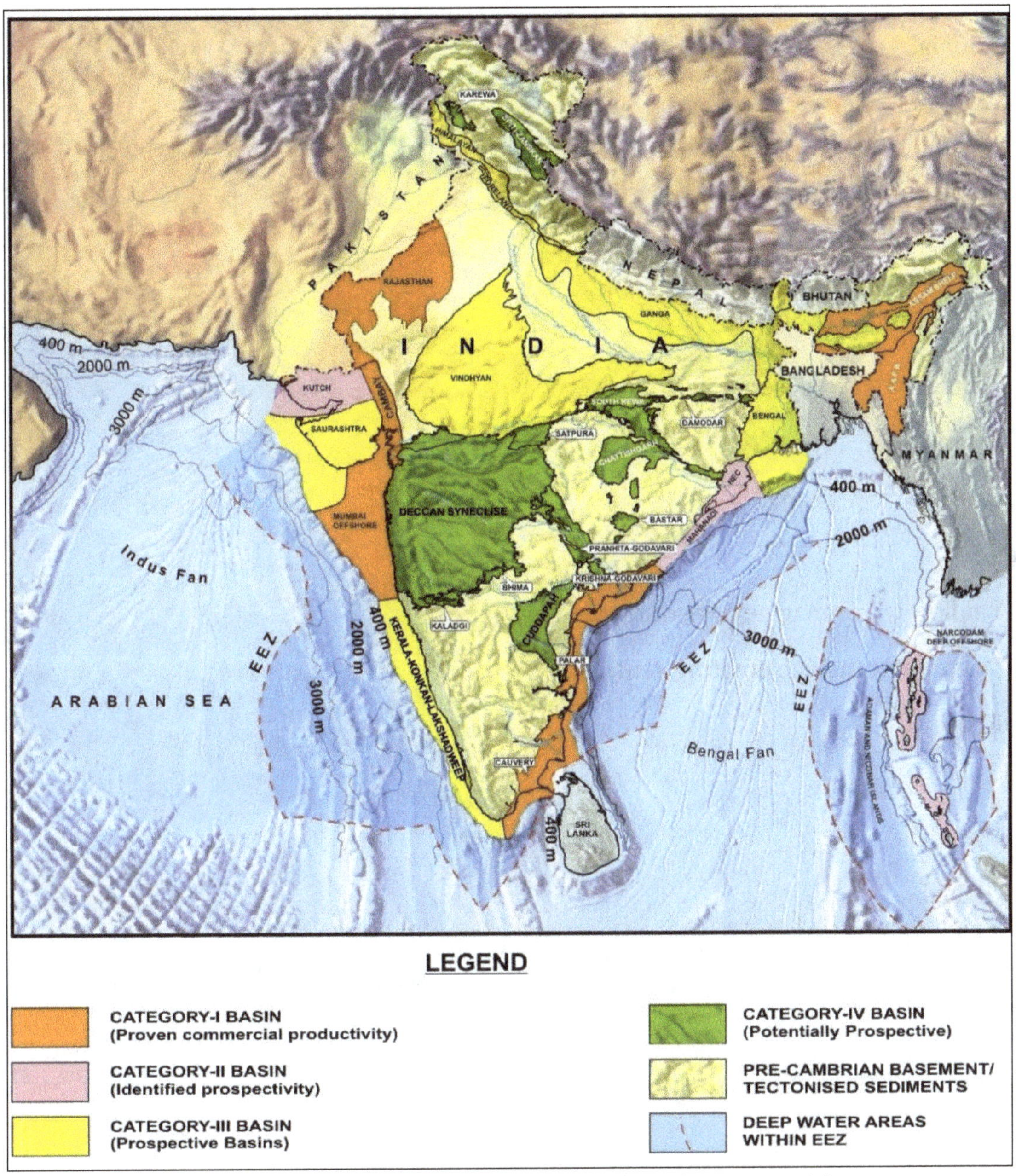

Sedimentary basins map of India and their hydrocarbon prospectivity.

Depositional Environment

Sedimentary depositional environment describes the combination of physical, chemical and biological processes associated with the deposition of a particular type of sediment and, therefore, the rock types that will be formed after lithification, if the sediment is preserved in the rock record. In most cases the environments associated with particular rock types or associations of rock types can be matched to existing analogues.

Types of Depositional Environments

Continental:

- Alluvial.
- Aeolian.
- Fluvial.
- Lacustrine.

Transitional:

- Deltaic.
- Tidal.
- Lagoonal.
- Beach.

Marine:

- Shallow water marine environment.
- Deep water marine environment.
- Reef.

Others:

- Evaporite.
- Glacial.

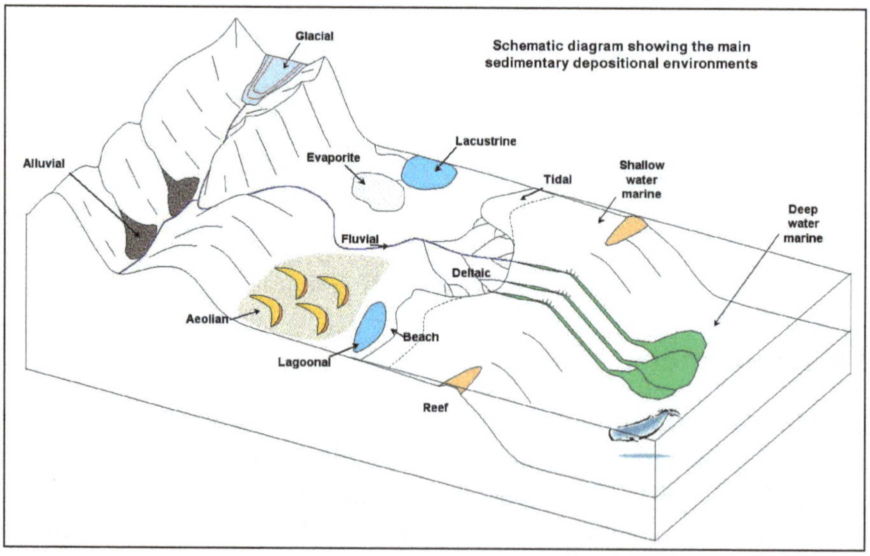

Schematic representation of different sedimentary depositional environments.

Depositional environments in older sediments are recognised using a combination of sedimentary facies, facies associations, sedimentary structures and fossils, particularly trace fossil assemblages, as they indicate the environment in which they lived.

Diagenesis

The physical and chemical changes that alter the characteristics of sediments after deposition are known as diagenesis. Some changes occur very near the surface at low temperatures and pressures while many changes require deeper burial (upto 5000m), several kilobars of pressure, and temperatures upto 300°. Diagenesis includes all those processes chemical, as well as, physical which affect the sediment after deposition up to the lowest grade of metamorphism (green-schist facies).

The early stage of diagenesis in terrigenous clastic rocks is mainly characterised by burial compaction as overburden pressure expels water from the pore spaces and reduces the porosity. Early cementation is found in particular environments otherwise it is not common. Like beaches where there is precipitation of calcite from seawater washing over the upper parts of a gravelly beach; calcrete, silcrete and ferricrete formation in soils, carbonate cementation forming hard grounds where there are very low rates of sea-floor sedimentation and gypsum cementation in the coastal plains of arid coasts.

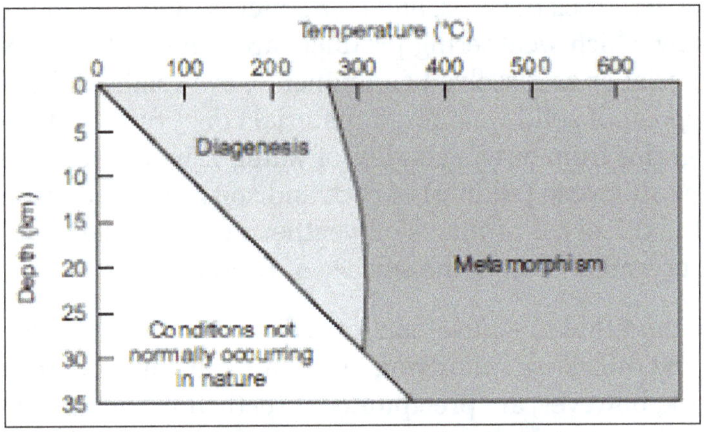

Depth and temperature ranges of diagenetic changes.

Mesogenetic cements are much more extensive in most clastic rocks in which authigenic minerals such as quartz, calcite and clays are present. Calcite is important cement in many sandstones and conglomerates deposited in marine environments. Quartz cements in sandstones commonly occur as syntaxial overgrowths and form adjacent to pressure solution contacts by diffusion along grain boundaries. Silica cements are commonly found where there is an absence of calcium carbonate, for example in quartz-rich sands deposited in a continental environment.

Economic Importance

Sedimentary rocks are the most common rock type found on the planet comprising nearly 70 percent of all rocks. These rocks from unlocking historical secrets to providing the power needed to sustain the modern world are critical to mankind and have been throughout history. Geologists usually find the outcrops on the surface of the earth to understand the geology but at the same time sub-surface geology to untap the hidden potential is also of vital importance. Exploration of mineral resources starts from surface investigation to deep mining and drilling. Important mineral resources to sustain the economy and civilization are found in sedimentary rocks. Resources like petroleum and natural gas, coal, uranium, iron, manganese, aluminium etc. are being explored and exploited for the decades. Sedimentology is one of the important tools to explore hydrocarbons and its relationship with petroleum industry is very productive. The study of sedimentary rocks and rock strata provides information about the subsurface that is useful for civil engineering, for example in the construction of roads, houses, tunnels, canals or other constructions.

Sedimentary Rock

Sedimentary rock is a rock formed at or near Earth's surface by the accumulation and lithification of sediment (detrital rock) or by the precipitation from solution at normal surface temperatures (chemical rock). Sedimentary rocks are the most common rocks exposed on Earth's surface but are only a minor constituent of the entire crust, which is dominated by igneous and metamorphic rocks.

Sedimentary rocks are produced by the weathering of preexisting rocks and the subsequent transportation and deposition of the weathering products. Weathering refers to the various processes of physical disintegration and chemical decomposition that occur when rocks at Earth's surface are exposed to the atmosphere (mainly in the form of rainfall) and the hydrosphere. These processes produce soil, unconsolidated rock detritus, and components dissolved in groundwater and runoff. Erosion is the process by which weathering products are transported away from the weathering site, either as solid material or as dissolved components, eventually to be deposited as sediment. Any unconsolidated deposit of solid weathered material constitutes sediment. It can form as the result of deposition of grains from moving bodies of water or wind, from the melting of glacial ice, and from the downslope slumping (sliding) of rock and soil masses in response to gravity, as well as by precipitation of the dissolved products of weathering under the conditions of low temperature and pressure that prevail at or near the surface of Earth.

Sedimentary rocks are the lithified equivalents of sediments. They typically are produced by cementing, compacting, and otherwise solidifying preexisting unconsolidated sediments. Some varieties of sedimentary rock, however, are precipitated directly into their solid sedimentary form and exhibit no intervening existence as sediment. Organic reefs and bedded evaporites are examples of such rocks. Because the processes of physical (mechanical) weathering and chemical weathering are significantly different, they generate markedly distinct products and two fundamentally different kinds of sediment and sedimentary rock: (1) terrigenous clastic sedimentary rocks and (2) allochemical and orthochemical sedimentary rocks.

Clastic terrigenous sedimentary rocks consist of rock and mineral grains, or clasts, of varying size, ranging from clay-, silt-, and sand-up to pebble-, cobble-, and boulder-size materials. These clasts are transported by gravity, mudflows, running water, glaciers, and wind and eventually are deposited in various settings (e.g., in desert dunes, on alluvial fans, across continental shelves, and in river deltas). Because the agents of transportation commonly sort out discrete particles by clast size, terrigenous clastic sedimentary rocks are further subdivided on the basis of average clast diameter. Coarse pebbles, cobbles, and boulder-size gravels lithify to form conglomerate and breccia; sand becomes sandstone; and silt and clay form siltstone, claystone, mudrock, and shale.

Chemical sedimentary rocks form by chemical and organic reprecipitation of the dissolved products of chemical weathering that are removed from the weathering site. Allochemical sedimentary rocks, such as many limestones and cherts, consist of solid precipitated nondetrital fragments (allochems) that undergo a brief history of transport and abrasion prior to deposition as nonterrigenous clasts. Examples are calcareous or siliceous shell fragments and oöids, which are concentrically, layered spherical grains of calcium carbonate. Orthochemical sedimentary rocks, on the other hand, consist of dissolved constituents that are directly precipitated as solid sedimentary rock and thus do not undergo transportation. Orthochemical sedimentary rocks include some limestones, bedded evaporite deposits of halite, gypsum, and anhydrite, and banded iron formations.

Sediments and sedimentary rocks are confined to Earth's crust, which is the thin, light outer solid skin of Earth ranging in thickness from 40–100 kilometres (25 to 62 miles) in the continental blocks to 4–10 kilometres in the ocean basins. Igneous and metamorphic rocks constitute the bulk of the crust. The total volume of sediment and sedimentary rocks can be either directly measured using exposed rock sequences, drill-hole data, and seismic profiles or indirectly estimated by comparing the chemistry of major sedimentary rock types to the overall chemistry of the crust from which they are weathered. Both methods indicate that Earth's sediment-sedimentary rock shell forms only about 5 percent by volume of the terrestrial crust, which in turn accounts for less than 1 percent of Earth's total volume. On the other hand, the area of outcrop and exposure of sediment and sedimentary rock comprises 75 percent of the land surface and well over 90 percent of the ocean basins and continental margins. In other words, 80–90 percent of the surface area of Earth is mantled with sediment or sedimentary rocks rather than with igneous or metamorphic varieties. The sediment-sedimentary rock shell forms only a thin superficial layer. The mean shell thickness in continental areas is 1.8 kilometres; the sediment shell in the ocean basins is roughly 0.3 kilometre. Rearranging this shell as a globally encircling layer (and depending on the raw estimates incorporated into the model), the shell thickness would be roughly 1–3 kilometres.

Despite the relatively insignificant volume of the sedimentary rock shell, not only are most rocks exposed at the terrestrial surface of the sedimentary variety, but many of the significant events in Earth history are most accurately dated and documented by analyzing and interpreting the sedimentary rock record instead of the more voluminous igneous and metamorphic rock record. When properly understood and interpreted, sedimentary rocks provide information on ancient geography, termed paleogeography. A map of the distribution of sediments that formed in shallow oceans along alluvial fans bordering rising mountains or in deep, subsiding ocean trenches will indicate past relationships between seas and landmasses. An accurate interpretation of paleogeography and depositional settings allows conclusions to be made about the evolution of mountain systems, continental blocks, and ocean basins, as well as about the origin and evolution of the atmosphere and hydrosphere. Sedimentary rocks contain the fossil record of ancient life-forms that enables the documentation of the evolutionary advancement from simple to complex organisms in the plant and animal kingdoms. Also, the study of the various folds or bends and breaks or faults in the strata of sedimentary rocks permits the structural geology or history of deformation to be ascertained.

Finally, it is appropriate to underscore the economic importance of sedimentary rocks. For example, they contain essentially the world's entire store of oil and natural gas, coal, phosphates, salt deposits, groundwater, and other natural resources.

Several subdisciplines of geology deal specifically with the analysis, interpretation, and origin of sediments and sedimentary rocks. Sedimentary petrology is the study of their occurrence, composition, texture, and other overall characteristics, while sedimentology emphasizes the processes by which sediments are transported and deposited. Sedimentary petrography involves the classification and study of sedimentary rocks using the petrographic microscope. Stratigraphy covers all aspects of sedimentary rocks, particularly from the perspective of their age and regional relationships as well as the correlation of sedimentary rocks in one region with sedimentary rock sequences elsewhere.

Classification Systems

In general, geologists have attempted to classify sedimentary rocks on a natural basis, but some schemes have genetic implications (i.e., knowledge of origin of a particular rock type is assumed), and many classifications reflect the philosophy, training, and experience of those who propound them.

The book Rocks and Rock Minerals by Louis V. Pirsson was first published in 1908 and it has enjoyed various revisions. Sedimentary rocks are classified there rather simplistically according to physical characteristics and composition into detrital and nondetrital rocks.

Numerous other attempts have been made to classify sedimentary rocks. The most significant advance occurred in 1948 with the publication in the Journal of Geology of three definitive articles by the American geologists Francis J. Pettijohn, Robert R. Shrock, and Paul D. Krynine. Their classifications provide the basis for all modern discussion of the subject. The nomenclature associated with several schemes of classifying clastic and nonclastic rocks will be discussed, but a rough division of sedimentary rocks based on chemical composition is shown in figure below:

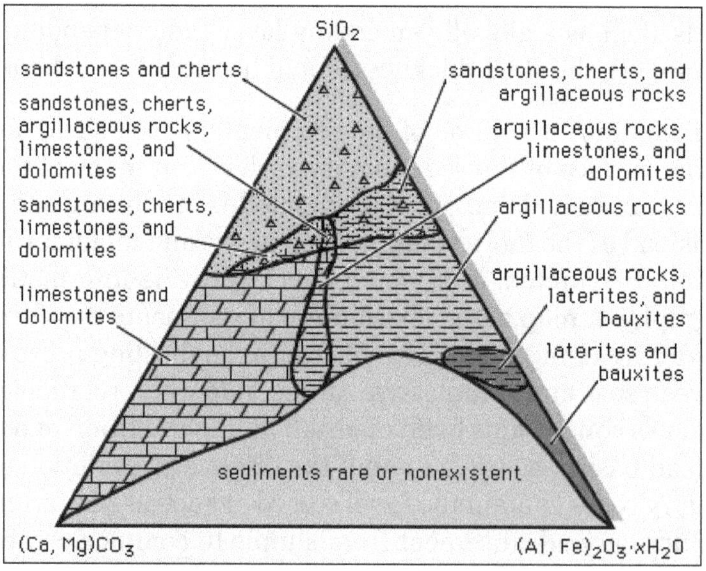

Chemical composition of sedimentary rocks.

Three major categories of sedimentary rocks are recognized: (1) terrigenous clastic sedimentary rocks, (2) carbonates (limestone and dolomite), and (3) noncarbonate chemical sedimentary rocks. Terrigenous clastic sedimentary rocks are composed of the detrital fragments of preexisting rocks and minerals and are conventionally considered to be equivalent to clastic sedimentary rocks in general. Because most of the clasts are rich in silica, they are also referred to as siliciclastic sedimentary rocks. Siliciclastics are further subdivided on the basis of clast diameter into conglomerate and breccia, sandstone, siltstone, and finer-than-silt-sized mudrock (shale, claystone, and mudstone). The carbonates, limestones and dolomites, consist of the minerals aragonite, calcite, and dolomite. They are chemical sedimentary rocks in the sense that they possess at least in part a crystalline, interlocking mosaic of precipitated carbonate mineral grains. However, because individual grains such as fossil shell fragments exist for some period of time as sedimentary clasts, similar to transported quartz or feldspar clasts, most carbonates bear some textural affinities to the terrigenous clastic sedimentary rocks. The noncarbonate chemical sedimentary rocks include several rock types that are uncommon in the sedimentary rock record but remain important either from an economic point of view or because their deposition requires unusual settings. Specific varieties include siliceous rocks (cherts), phosphate rocks (phosphorites), evaporites, iron-rich sedimentary rocks (iron formations and ironstones), and organic-rich (carbonaceous) deposits in sedimentary rocks (coal, oil shale, and petroleum).

Despite the diversity of sedimentary rocks, direct measurement of the relative abundance of the specific types based on the study of exposed sequences suggests that only three varieties account for the bulk of all sedimentary rocks: mudrock, 47 percent; sandstone, 31 percent; and carbonate, 22 percent. Another method, which involves comparing the chemical composition of major sedimentary rock types with the chemistry of Earth's continental crust, yields somewhat different numbers: mudrock, 79 percent; sandstone, 13 percent; and carbonate, 8 percent. Most sedimentary petrologists concede that the sedimentary rock record preserved and exposed within the continental blocks is selectively biased in favour of shallow-water carbonates and sandstones. Mudrocks are preferentially transported to the ocean basins. Consequently, indirect estimates based on chemical arguments are probably more accurate.

Terrigenous Clastic Rocks

A prominent physical feature of terrigenous clastic rocks is texture—that is, the size, shape, and arrangement of the constituent grains. These rocks have a fragmental texture: discrete grains are in tangential contact with one another. Terrigenous clastic sedimentary rocks are further subdivided on the basis of the mean grain diameter that characterizes most fragments, using the generally accepted size limits. Granules, pebbles, cobbles, boulders, and blocks constitute the coarse clastic sediments; sand-size (arenaceous) clasts are considered medium clastic sediments; and fine clastics sediments consists of silt- and clay-size materials.

The simplest way of classifying coarse clastic sedimentary rocks is to name the rock and include a brief description of its particular characteristics. Conglomerates and breccias differ from one another only in clast angularity. The former consist of abraded, somewhat rounded, coarse clasts, whereas the latter contain angular, coarse clasts. Thus, a pebble conglomerate is a coarse clastic sedimentary rock whose discrete particles are rounded and range from 4 to 64 millimetres (0.2 to 2.5 inches) in diameter. A more precise description reveals the rock types of the mineral fragments that compose the conglomerate—for example, a granite-gneiss pebble conglomerate.

Sandstones have long intrigued geologists because they are well exposed, are abundant in the geologic record, and provide an enormous amount of information about depositional setting and origin. Most schemes emphasize the relative abundance of sand-size quartz, feldspar, and rock fragment components, as well as the nature of the material housed between this sand-size "framework" fractions.

Fine clastics are commonly, but rather simplistically, referred to as mudrocks. Mudrocks actually can include any clastic sedimentary rock in which the bulk of the clasts have diameters finer than $^1/_{16}$ millimetre. Varieties include siltstone (average grain size between $^1/_{16}$ and $^1/_{256}$ millimetre) and claystone (discrete particles are mostly finer than $^1/_{256}$ millimetre). Mud is a mixture of silt- and clay-size material, and mudrock is its indurated product. Shale is any fine clastic sedimentary rock that exhibits fissility, which is the ability to break into thin slabs along narrowly spaced planes parallel to the layers of stratification. Despite the great abundance of the fine clastics, disagreement exists as to what classification schemes are most useful for them, and an understanding of their origin is hindered by analytical complexities.

Carbonate Rocks: Limestones and Dolomites

Limestones and dolostones (dolomites) make up the bulk of the nonterrigenous sedimentary rocks. Limestones are for the most part primary carbonate rocks. They consist of 50 percent or more calcite and aragonite (both $CaCO_3$). Dolomites are mainly produced by the secondary alteration or replacement of limestones; i.e., the mineral dolomite [$CaMg(CO_3)_2$] replaces the calcite and aragonite minerals in limestones during diagenesis. A number of different classification schemes have been proposed for carbonates, and the many categories of limestones and dolomites in the geologic record represent a large variety of depositional settings.

Noncarbonate Chemical Sedimentary Rocks

Noncarbonate chemical sedimentary rocks differ in many respects from carbonate sedimentary rocks and terrigenous clastic sedimentary rocks, and there is no single classification that has been universally accepted. This is a reflection of the great variation in mineral composition, texture, and other properties of these rock types. Such rocks as ironstones and banded iron formations (limonite, goethite, hematite, siderite, and chamosite), phosphorites, evaporites (rock salt, gypsum, and other salts), siliceous rocks (cherts), and organic-rich (carbonaceous) deposits of oil, natural gas, and coal in sedimentary rocks occur in much less abundance than carbonates and siliciclastic sedimentary rocks, although they may form thick and widespread deposits.

Classification schemes that incorporate all types of noncarbonate chemical sedimentary rocks do not exist because no triangular or tetrahedral scheme can accommodate all of them. Several of the major types are shown in the tetrahedron in figure below:

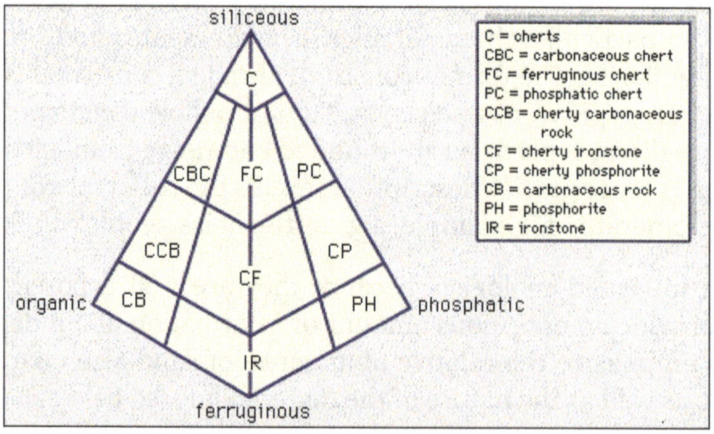

Noncarbonate, nonclastic sedimentary rocks: Common noncarbonate, nonclastic sedimentary rocks.

Sediment Gravity Flow

The term sediment gravity flow was introduced in geological literature to describe the major flow types involved in resedimentation processes. These flows have been defined and classified according to the dominant sediment support mechanism and rheology. The importance of clear idea about these flow types in understanding the subaqueous depositional systems has long been recognized, and during the last four decades, sediment gravity flow has emerged as a topic of special interest. Despite significant advancement in this field of research, some basic problems still persist in understanding these flow types. Critical analysis of the flow mechanisms reveals that the flow types commonly defined as sediment gravity flows markedly depart from the definitive properties of the latter. Certain contradictions in the proposed classifications of sediment gravity flow, based on rheology, make it more difficult, particularly for the beginners, to understand these high-density flows. Different views regarding characterization of these flow types have added a new dimension to this problem.

Middleton and Hampton defined sediment gravity flow as the flow of sediments or sediment–fluid mixtures under the action of gravity. According to them, sediment gravity flow is distinguished from fluid gravity flow by relative importance of sediment and fluid in driving the flow. They further emphasized that in a fluid gravity flow, the fluid is moved by gravity and drives the sediment along, but in sediment gravity flow, the sediments move under the action of gravity, and sediment motion makes the interstitial fluid moving. Middleton and Hampton recognized four distinct types of sediment gravity flow based on dominant sediment-support mechanism:

- Turbidity current, in which the sediment is supported by the upward component of fluid turbulence.
- Fluidized sediment flow, in which the sediment is supported by the upward flow of fluid escaping from between the grains as the grains are settled out by gravity.
- Grain flow, in which the sediment is supported by direct grain-to-grain interactions.
- Debris flow, in which the sediment is supported by a matrix (mixture of interstitial fluid and fine sediments) having finite yield strength.

Turbidity Current

The turbidity current is perhaps the most common and widely studied variety among the four flow types identified by Middleton and Hampton. Walker made an elaborate discussion on the historical development of the concept of turbidity current. Extensive laboratory studies have made a clear idea about the mechanism of turbidity current. It seems to be a special variety of gravity or density current, i.e., the flow of one fluid within another caused by the density contrast between them. Turbidity current is characterized by the flow of sediment–fluid mixture within an ambient fluid body, and the density contrast is caused by high concentration of sediments within the flowing mass. So far, as the densities of the fluid component of a turbidity current and the ambient fluid are concerned, these are likely to be the same; that is, the condition is expected to be homopycnal. In case of hypopycnal flows (fluid component of the flow is less dense), sediments get dispersed within the ambient fluid as buoyantly supported surface plume, and overflows set in (as in the case of river water charged with suspension cloud entering into a sea or hypersaline lake).

Experimental studies by Parsons et al. revealed that hyperpycnal flows might be generated from riverine outflows due to convective instability of the original hypopycnal plume. The term hyperpycnal, however, is used with reference to the relative density between the flowing sediment–water mixture and the ambient water body and not in the sense the term is being used here. Hyperpycnal condition (fluid component of the flow is denser) does not commonly occur in nature. There is no evidence of hypersaline flow entering into a freshwater body, and the temperature variation required for maintenance of requisite density contrast between these two fluids is not also expected at the site of mixing. Theoretically, however, hyperpycnal condition favors generation of turbidity current with much less grain concentration as compared to the homopycnal condition. The experimental observations depict that the grains are sustained within turbidity current due to the turbulent viscosity of the fluid component. Hence, under steady-flow condition, the fluid plays the dominant role in driving the flow.

Fluidized Flow

Sudden loss of shear resistance of sediments, in response to increase in pore-fluid pressure, may cause a change in state of the material from solidlike to liquidlike, and the process is commonly referred to as liquidization. Cohesiveness provides additional strength to the sediment to withstand such failure, and liquidization commonly takes place in cohesionless sediments.

Allen made an elaborate discussion on the process of liquidization, and he clearly pointed out that two distinct mechanisms can bring about such changes within the sediment. In liquefaction, the enhanced pore-fluid pressure causes total collapse of the grain fabric; the sediments lose their mutual contact and get dispersed within the fluid medium. Neither an inflow of pore fluid nor a change of volume of the deposit is involved in liquefaction. In case of fluidization, the fluid is forced upward through the grain mass until the immersed weight of the particles acting downward is balanced by the fluid drag. Influx of external fluid into the system is essential for fluidization.

Hence, the particle-support mechanism, assumed by Middleton and Hampton for the fluidized flow, actually corresponds to the liquefied character of the sediment. Lowe, however, considered liquefied and fluidized flows as separate types of sediment gravity flow with distinct particle support mechanism. Whatsoever, the enhanced pore-fluid pressure, responsible for the dispersion of the particles, is the main governing factor for generation of liquidized flows.

Grain Flow

The concept of grain flow was developed by Bagnold, who evaluated the dilatation within flowing mass of cohesion less particles due to gravity shearing. Lowe defined grain flow as the gravity flow of cohesion fewer solids maintained in dispersed state against the force of gravity by an intergranular dispersive pressure arising from grain interaction within the shearing sediments. In a grain flow, the interstitial fluid should be the same as the ambient fluid through which the flow is moving. If the dense interstitial fluid, overlying current or excess pore-fluid pressure aids significantly in maintaining the dispersion of a grain flow, the flow is designated as a modified grain flow.

Debris Flow

The concept of debris flow originally evolved from the subaerial sluggish flow of mixtures of granular solids, clay and water and was extrapolated to the subaqueous condition. Middleton and Hampton were of the opinion that in a 'true debris flow', the fluid, represented by clay–water mixture, has a finite cohesive strength, which provides a major support for the coarser grains, and this property distinguishes debris flow from other high-density flow types. Considering cohesiveness as the primary criterion for the characterization of a debris flow, Lowe redesignated the 'true debris flow' sensu Middleton and Hampton as cohesive debris flow.

Departure from the Definitive Properties

In sediment gravity flow, according to the original definition, the sediments are moved under the action of gravity, and the moving mass of sediments makes the interstitial fluid moving. But the flow types, identified as sediment gravity flows by Middleton and Hampton and Lowe, do not conform to the definitive properties of a sediment gravity flow for the following reasons:

- In major flow types, i.e., turbidity current, liquidized (both liquefied and fluidized) flows and grain flow, the particles remain in dispersed state, and the intergranular spaces cannot be maintained. Hence, the interstitial fluid becomes unconfined. The driving of interstitial fluid by the dispersed particles thus appears to be an unrealistic postulation.
- In turbidity current, the particles constitute relatively minor (because it is <50% by volume of the total moving mass) fraction of the flowing mass and remain in dispersed state within the turbulent fluid. The minor fraction of a moving system cannot make the major component moving, particularly when the latter acts as the host. The mechanisms of grain size differentiation of the particles within a turbidity current, as demonstrated by Middleton and Hand, further suggest that in this flowing system, the fluid component enjoys the freedom to behave independently and to control the movement of the grains within the flow.
- In turbidity current, the fluid turbulence is the main grain-support mechanism and is responsible for the maintenance of the flow. In liquidized flow, the increased pore-fluid pressure plays the same role, and the moving mass freezes (if not otherwise transformed) as soon as the pore-fluid pressure dissipates. Hence, in these two cases, fluid component plays the key role in driving the flow, and not the particles as assumed in the definition of sediment gravity flow by Middleton and Hampton.
- In a grain flow (or modified grain flow), as the particles are dispersed due to their mutual collision, the original configuration of the interstices cannot be maintained, and the fluid cannot also be retained there within during steady-flow condition. During the process of dispersion, the interstitial fluid thus becomes unconfined, and it interacts independently with the ambient fluid without any influence of the moving grains. In fact, the fluid does not play any role in driving a pure grain flow, but in modified grain flow, it does have some indirect influence. However, the idea of driving of the interstitial fluid by the moving grains does not appear valid in the case of grain flow also.
- In debris flow, the plastic character of flow suggests that the sediment–fluid mixture move en masse and neither of these two components plays the dominant role. However, if the

matrix and the coarser clasts are considered separately, as was done by Middleton and Hampton, the yield strength of the matrix material becomes the key factor in emplacement of a debris flow. The proportion of coarser particles is not likely to be sufficient to maintain their mutual contacts and the larger clasts float within the plastic fluid. Hence, the larger clasts do not make the fluid moving. The fluid, on the other hand, being sufficiently viscous, inhibits settling of the coarser clasts, and thus plays the governing role in maintaining the flow. Thus, the coarser particles in no way act as the sole driving factor for emplacement of debris flow.

Hence, the basic premise of the concept of sediment gravity flow appears fallacious. In fact, the flow types classified as sediment gravity flow are different enough from each other to be considered under a single category based on flow mechanism. Moreover, natural flows are emplaced in consequence of gravitational pull along the slope. Hence, the term 'gravity flow', as used by Middleton and Hampton to define sediment gravity flow and fluid gravity flow, is superfluous. The term gravity flow is often used as a synonym for density flow, where it implies gravitational separation of materials of different densities. Hence, the use of the term density flow sensu Mulder and Alexander may also appear confusing. However, to overcome these ambiguities, it is better to designate the sediment gravity flows as high-density flows to distinguish them from common surface flows involved in primary sedimentation processes.

Lowe made an attempt to classify the sediment gravity flows on the basis of rheological properties, and he opined that the flow character of low-density turbidity current varies from laminar to turbulent. But the natural flows are classified into two broad types—laminar and turbulent—on the basis of Reynolds number of the flow, and turbidity current, by definition, is a turbulent flow. Thus, identification of turbidity current as laminar flow is self-contradictory. Lowe and Postma considered grain flow as laminar flow. But Reynolds clearly pointed out that in a laminar flow, streamline motion prevails, and the fluids in adjacent layers mix only on a molecular scale. Hence, major interlaminar mass transfer and differentiation of sediments within flow are not expected in a laminar flow. On the contrary, in a grain flow, due to dominance of dispersive pressure, continuous mass transfer takes place in the direction at high angle to that of the flow. Differentiation of particles during flow also leads to the development of inverse grading in the resultant deposit. These characters of a grain flow markedly depart from the laminar flow character, and identification of grain flow as a laminar flow does not hold true. In case of liquidized flows, the high pore-fluid pressure causes upward movement of the pore fluid, and the movement of the liquidized mass cannot be confined to streamline motion. Hence, identification of liquidized flows as laminar ones also appears fallacious. In fact, most of the natural flows are turbulent in nature. In case of surface flows of Newtonian fluids, the boundary effect inhibits prevalence of laminar character. Only the high viscous flows (of non-Newtonian character) behave as laminar flow until the inertial flow forces exceed the optimum ratio with the viscous ones.

Problems Related to Characterization of Flow Types

Different approaches to characterize the high-density flow types, in most cases, have effectively made it more ambiguous. Shanmugam and Mulder and Alexander rightly pointed out that the same descriptive terms using various qualifications have been used to define different processes or products. Possibly because of receiving maximum attention, the turbidity current and the inferred turbidites have become the main areas of controversy. The 'turbidite mindset' guided a number of workers to apply the term turbidite to sediments deposited from flow types (mostly debris flows) radically different from the turbidity current sensu stricto. Shanmugam made an elaborate discussion on the extreme cases of such contradictions and advocated that the main reason behind these problems was the misapplication of the term 'turbidite' just to fit the studied deposits to some preassigned submarine fan models.

Kneller and Buckee raised a basic question on the concept of turbidity current itself and expressed their reservation regarding the way the term turbidity current is traditionally defined. They were of the opinion that the term turbidity current was adopted to describe a natural phenomenon whose exact nature is often unclear. It was also emphasized that in turbidity current, the sediment-support mechanism(s) is commonly impossible to determine in nature because a variety of sediment-support mechanisms may operate simultaneously within a sediment gravity flow. The history of development of the concept of turbidity current, however, does not indicate that the term was introduced 'to describe a natural phenomenon whose exact nature is often unclear' as has been claimed by Kneller and Buckee. The comments of Kneller and Buckee on sediment support mechanism further reflect their confusion between turbidity current and sediment gravity flow.

They also emphasized that the terms turbulent (disturbed by eddies) and turbid (opaque with sediment) have not been clearly distinguished in the geological literature, and this led to confusion in defining the flow mechanisms. The term 'nonturbulent turbidity current' of McCave and Jones, described as an oxymoron by Shanmugam thus received a strong favor. But there is a basic difference between turbulent flow and turbidity current. The turbidity current by definition is a turbulent flow, but all the turbulent flows (e.g., subaerial riverine flows) are not turbidity currents. Turbidity current is distinguished from other turbulent flows for its emplacement through an ambient fluid body. Moreover, the definition of turbidity current with reference to the sediment-support mechanism remained remarkably consistent over the past four decades. Hence, the doubt expressed by Kneller and Buckee on the possibility of determination of sediment-support mechanism in turbidity current does not hold true. Kneller and Buckee also introduced a term suspension current defined as 'flow induced by the action of gravity upon a (fluidal) turbid mixture of fluid and (suspended) sediment, by virtue of density difference between the mixture and the ambient fluid'. This definition, however, is not much different from that of the turbidity current, and as a result, the introduction of this new term 'suspension current' creates additional confusion.

Mulder and Alexander in their proposed classification of turbidity current introduced a term 'hyperpycnal turbidity current'. The adjective 'hyperpycnal' was used to mean that the bulk density of the effluent (sediment–water mixture) is greater than the receiving ambient water. But this is the essential condition for the generation and existence of turbidity currents. Hence, use of the term hyperpycnal to qualify a type of turbidity current is not only superfluous, but to some extent, also creates ambiguity in understanding turbidity current.

Attempts have been made by a number of workers to classify the turbidity currents into high-density and low-density types based on grain concentration. But the boundary between these two varieties has been set at different levels of grain concentration by different authors. However, Bagnold demonstrated that the gravity-driven turbulent currents could be maintained only for very low grain concentration (< 9% by volume). If the grain concentration exceeds this limit, according to Kneller and Branney, the fluid turbulence fails to support all the grains in suspension and grain-to-grain interaction becomes an increasingly important support mechanism. Hence, the flow can no longer be a turbidity current sensu stricto. Middleton observed a qualitative change in the behavior of the suspension at grain concentration of about 30%.

Shanmugam emphasized that in order to define high-density and low-density turbidity currents, there would have to be some defining concentration or density that would represent the bounds to these processes. But there is no such well-defined boundary. In alleviating certain conceptual problems, Shanmugam suggested to designate high-density turbidity currents with very high grain concentration as sandy debris flows.

The sandy debris flows, according to Shanmugam, represent a continuous spectrum of processes between cohesive and cohesionless debris flows. Commonly, they are rheologically plastic and occur with or without entrained turbulent cloud (like other debris flows) on top. The sediment-support mechanism includes matrix strength, dispersive pressure and buoyant lift. According to Middleton and Hampton, in a debris flow, grains are supported by the strength and buoyancy of the fluid medium rather than turbulence, upward escape of fluid or dispersive pressure. In contradiction to this postulation, Middleton and Hampton also admitted that dispersive pressure is significant in many debris flows. But the definitive properties, such as plastic rheology and laminar character of a debris flow, can no longer be maintained if dispersive pressure comes into existence as grain-support mechanism, and a transformation of flow will be the obvious consequence. Hence, the role of dispersive pressure as grain-support mechanism does not fit well into the rheology of a debris flow. Simple redesignation of high-density turbidity currents as sandy debris flow implies a sharp transition in rheologic property of a flow from Newtonian to plastic at a particular level of grain concentration. This appears to be an unrealistic postulation. In fact, the classification proposed by Middleton and Hampton and Shanmugam's proposition suffer from a common problem. In neither case, any flow type of intermediate character between turbidity current and debris flow was taken into consideration. This limitation, however, does not rule out existence of a sandy debris flow. Coussot and Meunier described debris flow as one-phase flow in which the whole mass undergoes continuous deformation and distinguished two extreme varieties—muddy debris flow (or cohesive debris flow) and granular debris flow (or noncohesive debris flow).

The validity of the proposition of the term sandy debris flow has also been questioned by Mulder and Alexander, who emphasized that from rheologic point of view, debris flows are unequivocally cohesive flows, and the flows involving noncohesive particles are frictional flows with very low cohesive force. But during characterization of cohesive flows, Mulder and Alexander seriously confused the relationship between cohesion and viscosity. The statement "cohesion and viscosity are not directly related" made by them not only goes against the general understanding about that relation but also contradicts the fact that "cohesion between particles results in high kinematic viscosity" quoted by them. It may be noted that viscosity is the modulus of shear deformation. It is commonly defined as the resistance to flow or shear deformation, and it depends on cohesion and rate of transfer of momentum between the flow laminae.

However, cohesion and friction between clastic grains, according to Dott, lend a degree of stability to sediments against the gravity-induced, downslope shear stress. Allen made an elaborate discussion on relative viscosity of sediment dispersion, and clearly stated that the relative viscosity increases rapidly with increasing particle concentration. It has also been pointed out that particle roughness and angularity significantly raise relative viscosity. Johnson, Lowe and Coussot and Meunier were also of the opinion that the debris flows freeze in consequence of frictional grain resistance (frictional freezing) and/or cohesive grain interactions (cohesive freezing). In fact, cohesive resistance and frictional resistance play the same role in shear deformation. In both of these cases, effectively, the viscosity increases with increase in shear resistance, and this, in turn, defines the rheologic property of the material undergoing deformation in the same way. Hence, in debris flows, the moving mass is characterized by finite yield strength due to either the cohesive strength of the material provided by high content of clay or frictional strength due to interlocking of grains. The whole mass keeps on moving until the shear force exerted by the downslope component of gravitational pull falls below the yield strength of the material and it freezes en masse.

Major and Iverson, however, opined that deposition from debris flow takes place primarily due to grain-contact friction and bed friction concentrated at the flow margins. Whatsoever, based on the nature of yield strength, the debris flows can be either cohesive or non-cohesive type. In cohesive debris flows, the plastic character develops due to high concentration of clay, and the clay–water mixture constitutes the fluid component of the flow. The high viscosity of the clay–water mixture provides support to the coarser sediments during steady-flow condition. It is noteworthy that the dynamic viscosity of clay–water mixture takes a sharp rise with increase in clay content beyond a certain limit. The variation diagram prepared by Grim depicts that for common native clays, this limit ranges between 22 and 30 wt.% of clay depending on the clay mineralogy (for attapulgite, Na-montmorillonite and Ca-montmorillonite, these values lie around only 5%). Thus, it may be assumed that a cohesive debris flow likely develops as the clay content exceeds this limit. The rheologically plastic behavior develops within cohesionless debris flows due to very high grain/water ratio. Kneller and Buckee theoretically deduced that a noncohesive particulate current with grain concentration of as high as 45% should rapidly become fully turbulent when the product of velocity and thickness exceeds about $3\times10^{-2}\ m^2\ s^{-1}$. Hence, a cohesionless debris flow is expected to be essentially sluggish and can be maintained only along gentle slope. Kneller and Buckee, however, made no specific mention about the nature of this flow transformation and expected mass transfer between the flow and the ambient fluid. The latter would change the concentration of grains within the flow. Calculations based on experimental data suggest that marine sediments with finest sand as the coarsest material can move as debris flow with bulk clay content as low as 2% (by weight), and the same with coarsest sand needs the bulk clay content of the order of 19% by weight. Experimental studies by Marr et al. revealed that some amount of clay is necessary to induce debris flowlike behavior in sandy dispersions moving under the influence of gravity.

The flow mechanics and resulting depositional features were found strongly tied to the coherence of the debris flows. Moderately and weakly coherent flows produced significant, low-concentration subsidiary turbidity currents, while the strongly coherent debris flows hydroplaned and generated only minor subsidiary turbidity currents. In a natural system, purely cohesionless dispersion is not normally expected. Small amount of clay impurity essential for maintenance of the coherent character remains in the system. According to Boswell, virtually all clastic sediments are, to some extent, cohesive in nature. Even saturated sand possesses a significant degree of bulk cohesion. In support of this view, Dott cited examples of contorted stratification in sandstones. Dott further emphasized that the ratio of surface area to mass of a particle chiefly determines if surface energy can produce significant adhesion between adjacent particles. Nevertheless, as long as the frictional resistance dominates over the cohesive resistance, the flow can be identified as cohesionless debris flow, and in cohesive debris flow, the cohesive material present within the system solely provides the shear resistance.

In an attempt to redesignate the flow types described as sandy debris flow by Shanmugam, Mulder and Alexander classified the noncohesive debris flows into hyperconcentrated density flows and concentrated density flows. They described these two varieties of density flows and turbidity current as subaqueous frictional flows. It was also mentioned that this approach distinctly differs from the idea of Kneller and Buckee; hence, it contradicts the general understanding about the mechanism of turbidity current because Kneller and Buckee grouped flows induced by the action of gravity upon a turbid mixture of fluid and (suspended) sediments by virtue of the density difference between the mixture and the ambient fluid as suspension current. So far as the mechanism of a turbidity current is concerned, it matches well with the suspension current described by Kneller and Buckee and in no way can be recognized as a frictional flow as postulated by Mulder and Alexander. In concentrated density flows, as proposed by Mulder and Alexander, grain-to-grain interaction generating dispersive pressure is a particularly important grain-support mechanism. Simultaneously, they classified grain flows under hyperconcentrated density flows. But the grain flows, by definition, are collisional flows characterized by dispersive pressure as the principal grain-support mechanism.

Definition of debris flow based on sediment-support mechanism may sometime cause problem in reconstruction of depositional mechanism from the resultant deposit. For example, the deposit from a plastic flow of a monodispersed system of clay-rich material will have no floating outsized clast. Obviously, the grain-support mechanism cannot be established due to the absence of coarser clast and this, in turn, will restrain to identify it as a debris flow deposit. On the other hand, for noncohesive debris flow, the effect of grain size on relative viscosity of sediment dispersion is less well understood. Still, experimental studies revealed reciprocal relationship between these two parameters. Thus, addition of outsized clasts in a monodispersed system may cause sharp fall in viscosity.

Thus, the cohesionless sediments deposited from a plastic flow may not contain any larger clast. On the other hand, the presence of large clasts within muddy matrix is not always an unequivocal evidence of deposition through en masse freezing of a cohesive debris flow because this can be attributed to many other depositional mechanisms. Large blocks, broken from submarine escarpments (commonly coastal cliffs or reef), can fall into deep water to accumulate as subaqueous taluses and get mixed up with deepwater, fine-grained sediments. Friedman et al. have elaborately discussed the processes involved in the origin of these sediments.

Postma et al. have experimentally demonstrated that in bipartite flows, outsized clasts can be transported along the rheological interface within the flow. The clasts glide along the top of a pseudolaminar inertia-flow layer, partly submerged in it, and are driven by the downflow component of turbulent shear stresses transmitted from the overlying, fast-moving turbulent layer. With deceleration, the inertia-flow phase of the flow freezes entirely and the large clasts are then trapped, suspended above the base of the resulting turbidite. The presence of large protruding clasts or subvertical clast alignment within cohesive matrix is an unequivocal evidence of emplacement of cohesive debris flow because these features indicate thixotropic nature of the transporting medium. But conspicuous absence of such clast orientations should not go against the interpretation as debris flow deposit because lack of such evidences and pervasive parallel-to-bedding clast alignment suggest a full laminar shear straining of the flow before deposition. Hence, much care has to be taken before drawing any inference about the grain-support mechanism and the flow mechanism accordingly for any fine-grained deposit with outsized clasts. Nemec and Steel have pointed out that the thickness of a conglomerate bed (BTh) and the bed's maximum particle size (MPS, defined as an arithmetic mean from 10 largest clasts) have often been used as an approximation of the thickness and competence of depositing flow at the point of measurement. Evidence from various sedimentary successions clearly indicates that the deposits of high-density flows display a significant, positive linear correlation between BTh and MPS. According to Nemec and Steel, it seems possible to infer from MPS/BTh diagram whether the debris flow conglomerate beds have been deposited from dominantly cohesive or cohesionless flows. They further emphasized that it is preferred to consider the diameter of the largest clast, rather than MPS, to speculate the magnitude of the cohesive strength of the flow from the grain size/bed thickness diagram because the largest clast size provides a better estimate of the flows competence. In the model proposed by Nemec and Steel, the presence or the absence of the ordinate intercept (indicating cohesive strength factor) of the best-fit regression line of grain size/bed thickness plots defines the nature of the debris flow.

Lowe specifically pointed out that flows that deposit through cohesive freezing encompass a broad spectrum of rheological behavior of sediment–fluid mixture, and the cohesive debris flows represent one end member. The resultant deposits from many of the cohesive flows show upper size limit of the coarser clasts embedded within clay-rich matrix. Larger clasts, if originally present, might settle through the matrix during flowage. The remaining clasts are uniformly dispersed within the cohesive matrix reflecting insignificant intergranular dispersive pressure and flow turbulence. Lowe further pointed out that in some of the cohesive flows, the larger clasts are not actually suspended within the clay–water mixture, but remain in mutual contact during rolling, sliding or intermittent bouncing downslope. The clay–water mixture provides buoyant lift and prevents frictional locking through lubrication. The resultant deposit shows clast-supported fabric with variable proportion of clay-rich matrix. According to Enos, many cohesive flows are turbulent at some stage and are capable to keep the coarser clasts in suspension. The deposition takes place through initial damping of turbulence and settling of coarser clasts followed by freezing of the remaining laminar phase.

The resultant deposit consists of a basal layer showing grain-supported fabric capped by a matrix-supported unit. Hence, identification of the wide spectrum of cohesive flows with varied flow characters, as debris flow cannot lead us to a specific definition of a debris flow. Debris flow is better to be identified as laminar flow of plastic material, which undergoes en masse freezing due to fall in shear force (downslope component of g) below the yield stress. Rheology, and not the sediment-support mechanism, should be the criterion for identification of a debris flow. The cohesive flows deviating from these essential characters have to be assigned separate status. Enos categorically emphasized that if laminar flow can be reasonably demonstrated in a medium, which was competent to transport large clasts, this may in turn be a strong argument for debris flow as opposed to current flow and possibly to other forms of subaqueous mass flow.

This may further be noted that a thick debris flow deposit is not commonly the product of abrupt freezing of an entire flow. Experimental and field observations suggest that a debris flow usually consists of a number of surges, and deposition takes place through incremental aggradation of individual surges. Each surge shoulders aside deposits of earlier surges, spreads out and leaves a deposit that is thinner and longer than the surge itself. Deposits of earlier surges do not consolidate before the arrival of the next surge, resulting in amalgamation of many surge layers into a single massive layer. Combined with the incremental aggradation, successive shouldering aside of earlier deposits by later surges prevents the protrusion of large clasts, which are therefore expected to be found mostly along the front and lateral margins of a deposit rather than on the top.

The concept of liquidized flow has an inherent problem. Liquidization, in general, comes into existence in response to sudden increase in pore-fluid pressure caused by some trigger. This causes instability within the sediments accumulated on a sloping surface, and the whole mass starts moving under the action of gravity. The initial instability is the prerequisite for the process of resedimentation, and liquidization is perhaps the most common cause of such instability. Once the unstable mass of sediment starts moving, the pore-fluid pressure tends to dissipate, and the flow continues until a stable fabric is restored. However, in common sedimentary flows, sustained liquidized flow is difficult to be conceived because discontinuation of the triggering mechanism inhibits maintenance of the high pore-fluid pressure. The pyroclastic flows often continue as liquidized flow for considerable time due to the process of self-liquidization caused by evolved gases under subaerial condition or steam generated from engulfed water under subaqueous condition. In case of ordinary sedimentary masses, such a process of self-liquidization is not possible. Middleton and Hampton also admitted that the main problem with fluidization as a sediment transport mechanism is that the excess pore-fluid pressure rapidly dissipates due to loss of pore fluid. The flow of sediment–fluid mixture initiated by the process of liquidization may either be transformed into a steady flow or the whole mass freezes as a massive bed through dissipation of the pore-fluid pressure. Major and Iverson experimentally observed that in debris flows, porefluid pressures, nearly sufficient to cause liquefaction, were developed and maintained during flow mobilization and acceleration, persisted within flow interiors during flow deceleration and deposition, and dissipated significantly only during consolidation of the sediments. In contrast, leading edges of debris flows exhibited little or no positive pore-fluid pressure.

Flow Transformation

Instability of the earlier sediments is the prerequisite for generation of flows involved in resedimentation processes. Such instability normally comes into existence either due to oversteepening of the parent deposit or through some process of liquidization including seismogenic slumping and storm wave-induced landslide. Evidence of remobilization of sediments through subaqueous sliding and slumping has also been reported by a number of workers. Friedman et al. specifically pointed out that subaqueous slumping is broadly of two types: (1) coherent slumping and (2) incoherent slumping. In the former case, the strata within the displaced mass remain intact and may be deformed, while in the latter, the entire body of sediments is thoroughly mixed and churned, and almost all traces of stratification are obliterated. The process of breaching can be another potential cause of instability within a sedimentary deposit.

Ideal grain flows, exemplified by the avalanching of loose sediments along the slip face of an Aeolian dune, are caused by over-steepening of the sediments accumulated between the summit and brink zone of the dune. Remobilization of subaqueous sediments caused by instability may lead to generation of debris flow or turbidity current. It has already been mentioned that the common sediments do not undergo the process of self-liquidization. Hence, sustained liquidized flow of sediment–fluid mixture is not expected. But after initial movement, if the downslope component of the gravitational pull exceeds the yield strength of the moving mass, the flow continues as a debris flow. Under subaqueous condition, the moving sediment–fluid mixture undergoes body transformation on reaching a steeper slope. Under such condition, the moving mass is accelerated due to enhanced gravitational pull, and the flow tends to become turbulent. This further disturbs the relative tranquility of the adjacent fluid due to inertia contrast, and a part thereof gets mixed up with the sediment–fluid mixture, and a turbidity current comes into existence due to decrease in grain/water ratio of the flowing mass.

Experimental observation also goes in favor of dilution of high-density flows through entrainment of ambient fluid into the flow head. Shanmugam clearly stated that on increase in fluid content, a plastic debris flow moving downslope might evolve into a fluidal turbulent flow—the turbidity current. Hampton postulated that dilution and stripping of surface materials might cause generation of turbidity currents from subaqueous debris flows. Piper et al. also predicted transformation of subaqueous debris flow into turbidity current. Penetration of water into the interior of a flow through clefts developed along the flow front, or detachment and disintegration of a hydroplaning flow front may lead to transformation of subaqueous debris flows to turbidity currents. Entrainment at the upper boundary of the body or by passage through a hydraulic jump also causes flow transformation. Lee et al. apprehended progressive dilution of a debris flow due to mixing of ambient water and removal of suspended sediment by deposition during downslope movement. Hallworth et al. also postulated that the transformation from one to the other flow types may be progressive as a result of entrainment and dilution. Two possible trends of such flow transformation have been suggested by Sohn et al. According to Sohn et al. the flow head of a debris flow, containing the densest slurry and the coarsest clast fraction, consumes much kinetic energy and, hence, moves slower than the main body of the flow. This causes piling up of slurry behind the flow head leading to longitudinally segregated flow comprising preceding debris flow and a trailing dilute flow. The reverse trend may also come into existence when the grain/water ratio of turbidity current keeps on increasing through incorporation of sediment during the process of erosional incision. An intermediate phase, generated in the process of flow transformation, may possess distinctive character to be identified as a separate flow type as exemplified by the hyperconcentrated flow. Pierson and Costa pointed out that the rheology of a sediment–fluid mixture is mainly governed by concentration of sediments, and to a lesser extent, by the size, physical and chemical properties of these transported particles. Shanmugam opined that high grain concentration tends to damp the turbulence, the very property that distinguishes turbidity currents from other high-density flows.

Beverage and Culbertson were of the opinion that flows with sediment content between 40% and 80% by weight (or 20–60% by volume) are intermediate in sediment transport characteristics between debris flow and normal stream flow, and they are designated as hyperconcentrated stream flows. Smith pointed out that strict application of this definition of hyperconcentrated stream flow to a depositional record is not possible because the definitive parameter, sediment concentration, cannot be measured in a deposit. Smith however, recommended the term hyperconcentrated flood flow to describe high-discharge flows in which neither turbulence is the lone sediment-support agent nor in which deposition occurs en masse. Sohn et al. made an elaborate discussion on hyperconcentrated flood flows and sorted out the following general properties of such flows from published literature:

- These flows have two-phase or multiphase character. The coarser clasts (sand and gravel) and the fluid component (mixture of water, silt and clay) behave independently.
- The fluid component possesses meager yield strength and acts as a transporting medium only.

- The role of turbulence may be important as a grainsupport mechanism.

However, existence of hyperconcentrated flood flow with its distinctive character is well recognized and, hence, can be considered as a distinct flow type. Lowe and Guy introduced the term slurry flow to refer to flows transitional between turbidity current and debris flow. In slurry flows, described by Lowe and Guy, the clast fraction, however, lies within sand-sized grains. Hence, in the present study, a more general term, hyperconcentrated flow, is used to designate the flow types having character intermediate between turbidity current and debris flow. This is a flow in which the particles are neither primarily supported by flow turbulence nor a plastic character develops within the flowing mass to cause en masse freezing.

Parsons et al. experimentally demonstrated the nature of transition between cohesive flow and grain flow behavior caused by flow transformation within debris flow. The cohesive flow character with finite yield strength and shear-thinning rheologies were retained within the body of the debris flow, while frictional behavior was observed at the flow front or snout. Hampton was also of the opinion that flow transformation may occur through mixing and dilution at the flow head. Gravity transformation causes segregation of coarser material at the lower part of the flow leading to development of a high-concentration underflow with an overriding dilute phase. Bipartite flow can also be generated by an excess of sediment supply toward the flow base relative to the deposition rate. However, the highly concentrated bed load layers, developed beneath and driven by overlying turbulent flows, are referred to as traction carpets. The traction carpet itself is again density stratified, and according to Sohn, its variable rheological properties cannot be explained by the uniformly dispersed grain flow model of Bagnold. Sohn suggested that a traction carpet can be divided into a lower frictional region of nonvibrating grains with continuous contacts, and an upper collisional region dominated by active grain collisions. In frictional region, generation of dispersive pressure is probably hampered by the relative low strain rate, almost continuous grain contacts due to very high particle concentration, and loading by the materials from overlying regions.

Turbulence in this region is almost suppressed, and the region behaves as non-Newtonian fluid with plastic rheology, very similar to that of a sandy debris flow as suggested by Shanmugam. The collisional region is subjected to intense shear stress exerted by the overlying turbulent flow and is characterized by active grain collision and generation of dispersive pressure, a situation qualitatively very similar to that of grain flow as defined by Lowe. The term modified grain flow is used to designate the subaqueous grain flows where the fluid significantly aids in maintaining the dispersion. Even if the subaqueous grain flow does not receive any such direct favor from the surrounding fluid, dispersion of immersed particles is still promoted by the buoyant force exerted by the fluid medium. In subaerial condition, avalanching of dry particles is free from any such partial influence of the fluid component. Hence, it is better to reserve the term grain flow exclusively for subaerial granular flows where the fluid component does not play any role, and the term subaqueous grain flow may be introduced to designate their subaqueous counterparts. Modified grain flow thus becomes a specified variety of subaqueous grain flows.

Classification of Natural Flows

The preceding discussion reveals that the flow types involved in resedimentation processes are commonly generated through transformation of flows. Flow transformation mainly comes into existence due to gradual change in grain/water ratio of the flowing system. This affects the rheology of the flowing mass, and a corresponding change in its response to gravitational pull sets in. The grain/water ratio can change gradually in either direction, and intermediate stages are likely to be developed.

Hence, each of the flow types described above, instead of being an independent system, is a well-defined segment of a continuous system. An empirical relationship between these flow types can be conceived with reference to the grain/water ratio of the flowing mass. This relation can be extended to bring all aqueous flows within a single frame between two end members—pure fluid flow and subaqueous grain flow. Pure fluid flow can be defined as the flow of fluid with no or negligible amount of transported particles. On the other end, the subaqueous grain flow represents flow of cohesionless solid particles, and water plays effectively no direct role in driving this flow. Addition of particles in pure fluid flow gives rise to sediment laden flows. These flows are characteristically turbulent in nature. The sediment-laden flows are broadly of two types based on the proportion of the solid particles in the flowing mixture. The low-density flows contain low amount of transported materials and commonly occur as surface flows; the best example of which is the riverine flow. Increase in the proportion of grains causes corresponding increase in the density of the sediment–fluid mixture. Subaerial flow of these high-density mixtures is exemplified by the mudflows. If the same is emplaced within ambient water body as density currents or underflows, turbidity current is generated. On further increase in grain/water ratio, the flow character changes to hyperconcentrated flow and then to debris flow. These two flow types are emplaced under both subaqueous and subaerial conditions.

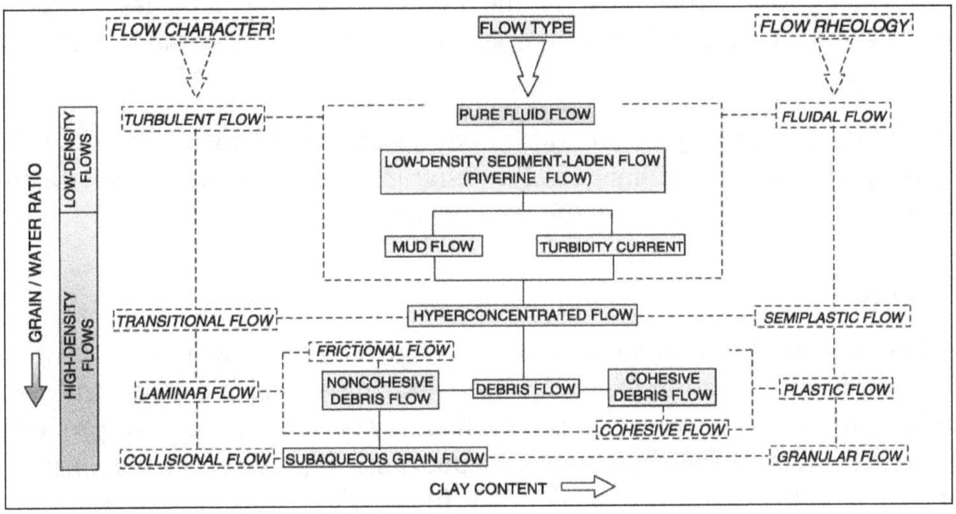

General classification of aqueous flows based on the trend of change in character and rheology of the flow with proportion and composition of the grains in the flowing system.

However, in either situation, water remains the sole fluid component. Definition of specific boundaries between these flow types is difficult to be ascertained with reference to the grain/water ratio of the flows. It may be assumed from experimental observations of Middleton that in turbidity currents, the maximum grain concentration must not exceed 30% by volume. On the other hand, the rheology of the flowing mass markedly changes with average clay content above approximately 27% (by weight). Hence, it may be empirically concluded that in turbidity current and hyperconcentrated flow, the grain population is mainly represented by noncohesive fraction with relatively small amount of clay. In noncohesive debris flow, the contribution of clay is likely to be much less. As the clay content reaches a value of approximately 27% by weight of the flowing mass, the flow is expected to behave as a cohesive debris flow irrespective of the proportion of noncohesive grains in the sediment–fluid mixture. In subaqueous grain flow, the grains and fluid move more or less independently, the effective grain/water ratio becomes very high and may be considered very close to that of pure grain flow where the fluid has practically no role in driving the flow. Liquidization may lead to initiation of most of the resedimentation processes, but the liquidized flows, as such, are short-lived ones and are related to some external triggers. Sustained liquidized flows of sediment– fluid mixtures are not expected in nature. Hence, this flow type has been excluded from the present scheme of flow classification.

The flow types described above can be classified under four broad categories based on flow rheology: (1) fluidal flow, (2) semiplastic flow, (3) plastic flow and (4) granular flow. Within fluidal flow, pure fluid flow, riverine flow, mudflow and turbidity current are characteristically turbulent in nature. The first two varieties are low-density flows, the third one is highdensity surface flow, and the last one is high-density underflow. These flows, having Reynolds number>2000, can be grouped together as turbulent flows. The hyperconcentrated flows of semiplastic rheology show transition character between turbulent flow and laminar flow and can be considered as transitional flow (Reynolds number lying between 500 and 2000). The plastic flow is represented by debris flows, which are of two types—the cohesive flow (cohesive debris flow) and frictional flow (noncohesive debris flow). Both of these varieties of plastic flow are characterized by their laminar behavior (Reynolds number < 500). The granular flow mainly includes the subaqueous grain flows and grain flows, which proceed through dynamic dilatation, and it can be designated as collisional flow. Avalanche, debris falls and subaqueous rock falls are the other possible varieties of granular flow.

Hence, the present classification gives an idea about the interrelationship between different types of aqueous flows and the nature of change in flow character and flow rheology with change in proportion and composition of the grains in the flowing system.

Sediment Budget

The sediment budget concept provides a valuable integrating framework for studying the various processes of sediment mobilization and delivery operating within a catchment. Most of the key processes involved and some of the recent research aimed at developing an improved understanding of these processes. Although the emphasis is commonly placed on the magnitude of the fluxes and stores, and thus on the quantities of sediment involved, it is also important to recognize that the properties of the sediment associated with different components of the budget may change, as sediment is transferred from source to sink. Mobilization, transfer, and deposition processes will frequently involve selectivity related to both particle size and particle density, resulting in contrasts between the composition of sources and sediment associated with different components of the budget. In the case of variations in particle size, the contrast between the effective or in situ grain size and the ultimate or absolute grain size of the sediment, which reflects the existence of composite particles (i.e., aggregates and flocs), can introduce further complexity and exert an important influence on the properties of sediment associated with individual components of the sediment budget.

In the case of sinks, for example, coarser particles may be preferentially deposited, but if these coarser particles comprise aggregates or flocs, they may contain a significant proportion of fine particles, the deposition of which might otherwise be unexpected. Droppo has called for a rethinking of conventional approaches to investigating suspended sediment dynamics to reflect the existence and importance of such composite particles. Equally, sediment mobilized from different sources may be characterized by different properties, and the sediment associated with different components of the budget may change according to its source or the relative contribution from different sources. In this context, contrasts in the properties of sediment derived from different sources may reflect both the source type (e.g., sheet and rill erosion vs. gully and channel bank erosion) and the spatial variability of source material properties caused by variations in geology, soil type, or land use across a catchment.

Key characteristics of the functioning of a sediment budget include its connectivity and thus the extent to which the slopes or upstream parts of a catchment are linked to the channel system or the catchment outlet. The connectivity of a system will depend on the incidence and efficiency of the transfer pathways and the magnitude of the stores. The connectivity of the system is clearly of fundamental importance when investigating and attempting to control sediment-induced diffuse source pollution. Detailed assessment of connectivity necessitates consideration of the transfer pathways involved and their efficiency in transferring sediment through the sediment delivery system. By providing a clearer representation of the links between erosion or sediment mobilization and sediment yield, the catchment sediment budget represents an important advance over the sediment delivery ratio concept. The latter simple black box concept recognized that the sediment output was likely to be less than the gross sediment mobilization and represented the ratio of the former to the latter. The magnitude of this ratio was in turn linked to the size of the catchment, with its magnitude commonly decreasing as the scale of the catchment increased. Many limitations of the sediment delivery ratio concept have been widely debated.

The precise form taken by the sediment budget of a catchment will reflect a wide range of controls, including the local topography and the hydrological regime, as well as the size of the catchment. It provides an indication of the potential nature and extent of such variability by indicating the key characteristics of the sediment budgets of four small drainage basins on the Russian Plain documented by Golosov et al. These are all relatively small basins, heavily impacted by agricultural land use and associated soil erosion. The investigation aimed to establish the proportion of the sediment mobilized within the catchments by different erosion processes that reached the basin outlets. In this environment, three key sediment mobilization processes were identified. These comprised sheet erosion (i.e., widespread erosion of the surface by surface wash), rill erosion (i.e., erosion by concentrated flow in micro-channels developed on the slopes), and gully erosion (i.e., erosion within deeper ephemeral channels that dissect the landscape and where sediment is mobilized by mass movements on the gully sides as well as by the flow through the gully). In this environment, sheet and rill erosion are generally more important than gully erosion as a sediment source and there is little evidence of sediment storage on the lower parts of the slopes. Slopes are frequently convex, terminating at the margins of balkas (flat floored, gully-like features), that dissect the landscape. Even within this relatively homogeneous area, the proportion of the sediment mobilized by erosion within the individual catchments that reaches the basin outlet ranges from 0% to 89%. In most of the catchments, both the balka bottoms and the river floodplains constitute major sinks for sediment moving through the system and as with Coon Creek, the sinks represent a very important component of their sediment budgets.

As the scale of the drainage basin increases, deposition of sediment on the river floodplains in the lower parts of the basin will commonly assume increasing importance. Work within the catchments of the Rivers Ouse (3315 km^2) and Wharfe (818 km^2) in Yorkshire, UK, reported by Walling et al. has, for example, shown that as much as 30–40% of the sediment delivered to the main channel system is deposited on the adjacent floodplains during overbank flood events and does not reach the basin outlet. At a larger scale, Bobrovitskaya et al. provided information on the sediment budget of the lower River Ob which drains a vast catchment of 2 950 000 km^2 in Siberia to the Arctic Ocean. The available information on the suspended load of this river provided by two gauging stations on its lower reaches separated by an 870-km reach indicates that in its lower reaches approximately 50% of its suspended sediment load is deposited on the well-developed floodplain bordering the river and fails to reach the lowest measuring station. At this larger scale, tectonic subsidence within the interior of a river basin can also promote the development of major sediment sinks which reduce the downstream sediment flux. This is well illustrated by the Rio Madeira, a major tributary of the Amazon in Bolivia and information. The upper basin of this river, which extends to ~170 000 km^2, drains the Andean Cordillera where erosion rates are high and consequently transports a very high annual suspended sediment load of ~500–600 Mt. However, on leaving the Andes, the downstream course of the Rio Madeira passes through a subsiding foreland basin where of the order of 270 Mt of sediment is deposited each year. As a result, only about 45% of the upstream load of the Rio Madeira is transported downstream into the main Amazon River system.

The operation of sediment budgets can be considered over several different timescales. Several recent sediment budget investigations that have taken a longer-term perspective have emphasized the importance of sediment storage, with the majority of the sediment mobilized being stored within the catchment over long periods. For example, Rommens et al. reported a Holocene sediment budget for a small 103 ha agricultural catchment in the Belgian loess belt that shows that 58–80% of the sediment mobilized within the catchment had been stored near its source and not delivered to downstream rivers. Similarly, Prosser et al estimated that as much as 80% of the sediment eroded from large coastal catchments in Eastern Australia in historical times remains stored in their channels and floodplains. Many of the sediment sinks associated with a sediment budget are likely to be long-term sinks. For example, the sediment deposited on the lower parts of a slope will commonly remain in near-permanent storage, unless there is a significant change in the pattern of erosion. River floodplains will frequently also represent longer-term sinks, with bank erosion perhaps causing some loss, which is balanced by point bar formation and deposition elsewhere. Floodplain sinks could, however, is rapidly remobilized by changes in channel pattern and increased channel migration associated with changes in the flow regime caused by human activity in the upstream catchment or climate change.

Some sinks will, however, operate as shorter-term stores. This was the case with the middle valley sink within the Coon Creek catchment depicted in figure below. Furthermore, at the annual timescale, sediment deposited within the channel system may accumulate during one period of the year, only to be remobilized and flushed out during a subsequent period. In this situation, storage is clearly temporary and in their study of three groundwater-dominated lowland catchments in the UK, Collins and Walling demonstrated that fine sediment accumulated within the channel during the winter period, when most sediment was transported through the system, and was subsequently remobilized during the summer period. Estimates of the average mass of sediment stored in the channel systems of the three catchments during the 2-year study period demonstrated that this was equivalent to between 21% and 38% of the mean annual suspended sediment export from the three catchments. By documenting changes in storage through time, it was also possible to estimate the total amount of sediment entering and leaving channel storage within the three catchments over the study period. The amounts of sediment entering and leaving channel storage within the three catchments were equivalent to between ~20% and 75% and between 25% and 70%, respectively, of the mean annual sediment yield, demonstrating that a significant proportion of the sediment flux passed through this short-term sink.

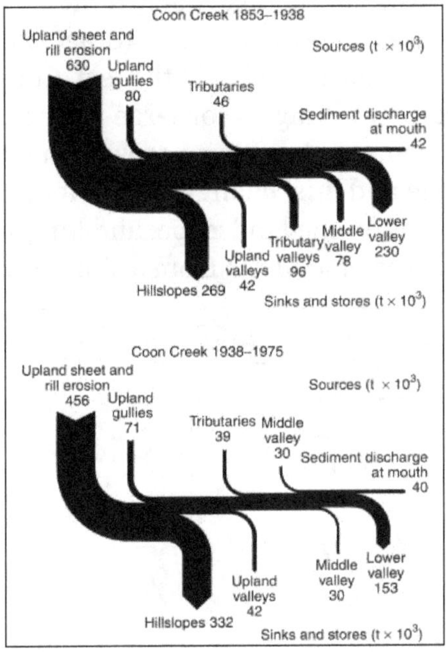

The sediment budgets for Coon Creek, Wisconsin, USA for the periods 1853–1938 and 1938–1975 produced by Trimble. The fluxes shown are mean annual values. From Trimble a sediment budget for Coon Creek basin in the Driftless Area.

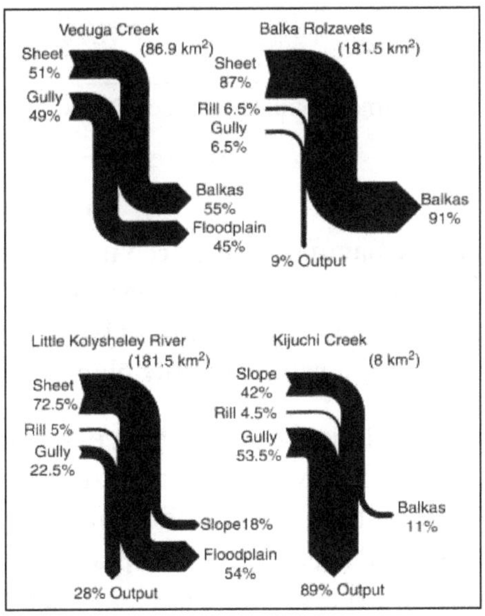

The sediment budgets for four small drainage basins on the Russian Plain, established by Golosov et al. From Golosov et al. Sediment budgets of river catchments and river channel aggradation on the Russian plain.

As with channel classification, sediment budgets are both an approach and a model. A sediment budget is a rigorous quantitative accounting of the production, transfer, and storage of sediment within a landscape. It is a quantification and grounding of the concept of the fluvial system in time and space. An early and influential quantitative sediment budget was developed by Dietrich and Dunne for a small coastal watershed in Oregon. They provided a conceptual framework and suggested a range of approaches for estimating fluxes of sediment by various processes through the landscape. The conceptual model was a flowchart with storage, fluxes, and outputs represented by geometric figures; estimates of rates were specified in the text. Coupled with new dating techniques, such as cosmogenic nuclides, sediment budgets for large river systems have been developed that demonstrate the role of episodic events in long-term erosion and continental shelf sedimentation, local variations in sediment storage and flux, changing contributions of different sediment sources over time, effects of human activities on the landscape, and long-term landscape evolution. This is another example of wholesale returns of analysis from a simple conceptual frame.

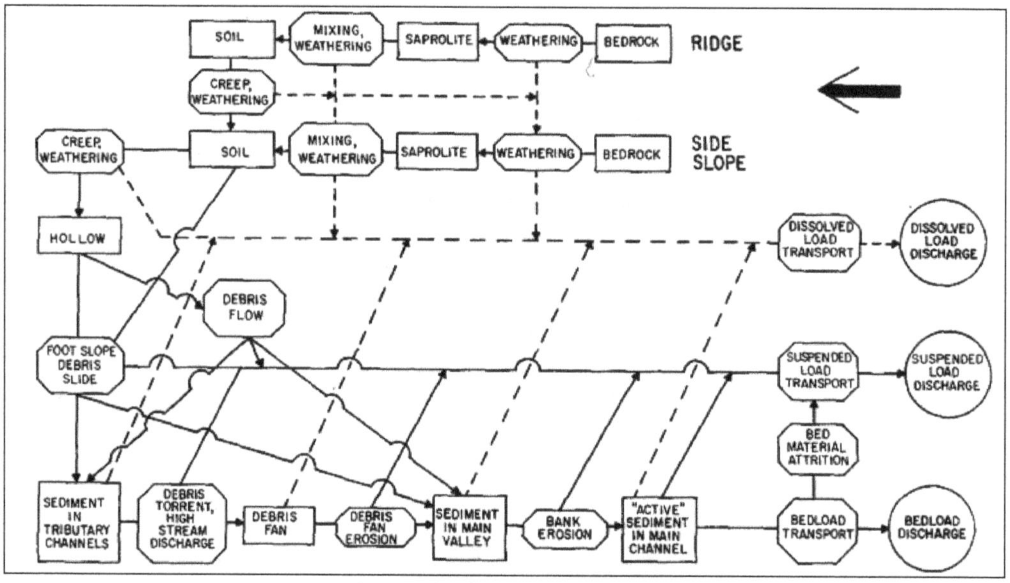

Sediment budget for Rock Creek.

Sedimentary Basin

The official definition of a sedimentary basin is: a low area in the Earth's crust, of tectonic origin, in which sediments accumulate. Sedimentary basins range in size from as small as hundreds of meters to large parts of ocean basins. The essential element of the concept is tectonic creation of relief, to provide both a source of sediment and a relatively low place for the deposition of that sediment.

Keep in mind that a sedimentary basin doesn't have to be a place on the Earth's surface with strictly basinal shape, with closed contours, like a washbowl: great masses of sediment can be deposited on a surface with a gentle and uniform slope. But implicit in the concept of a sedimentary basin is the existence of prolonged crustal subsidence, to make a place for a thick deposit of sediment that might well have been deposited in an area without basinal geometry at the surface. Tectonics is needed to make sedimentary basins, but the record of the basin itself is sedimentary.

As with most blanket statements, the one above has exceptions to it. A sedimentary basin can be made just by erecting high land in an adjacent area by volcanism.

The term "sedimentary basin" is usually not applied to relatively thin and very extensive deposits of sandstone, limestone, and shale from epicontinental seas on the cratons many of which have seen no deformation through billions of years, but only to relatively thick deposits in tectonically active areas with negative relief.

Paleocurrents

Much effort has gone into developing ways of figuring out paths of dispersal of sedimentary material in basins. One of the standard ways is to measure paleocurrent directions recorded locally in the rocks. (A paleocurrent is just what the term implies: a current, of water or wind that existed at some time in the past.) Techniques are well established.

Knowledge of paleocurrents is helpful in solving both local and regional problems of sedimentary basins. Locally, paleocurrent directions can help you to figure out or predict, indirectly, the shape and orientation of sediment bodies, like channel sandstones. This has obvious advantages in petroleum exploration. Regionally, paleocurrent directions can help establish paleoslope and source of sediment supply to the basin.

The paleocurrent measurements you take from dipping beds are no good in themselves: what you need to do is "undeform" the strata by rotating them back to horizontal, taking your paleocurrent measurements with them. That's straightforward (using a stereonet by hand or a computer program) provided that the strata are not strongly deformed. But the greater the deformation, the more uncertain is the exact way you should be undeforming the strata.

Basins Making Procedure

In one sense, the origin of sedimentary basins boils down to the question of how relief on the Earth is created.

Local

On a small scale, hundreds to thousands of meters laterally, fault movements can create relief of hundreds to thousands of meters, resulting in small but often deep basins (some of these are called intermontane basins; think about places like Death Valley). You might guess that it takes dip-slip fault movements to create new relief, but that's not true: steps (in the proper sense) along strike-slip faults can produce small pull-apart basins; more on them later. Relief of this kind is on such a small scale that it tends not to be isostatically compensated. It's like setting a block of granite out on your driveway; the flexural rigidity of your driveway is great enough compared with the imposed load that the granite block is prevented from finding its buoyant equilibrium position.

Regional

Basin relief can be created mechanically on a regional scale in two very important ways: thermally or flexurally, or by a combination of those two effects). Keep in mind that basins can also be made just by making mountain ranges, on land or in the ocean, by volcanism.

Thermal

If the lithosphere is heated from below, it expands slightly and thus becomes less dense. This less dense lithosphere adjusts isostatically to float higher in the asthenosphere, producing what we see at the Earth's surface as crustal uplift. If the lithosphere cools back to its original temperature, there's isostatic subsidence back to the original level.

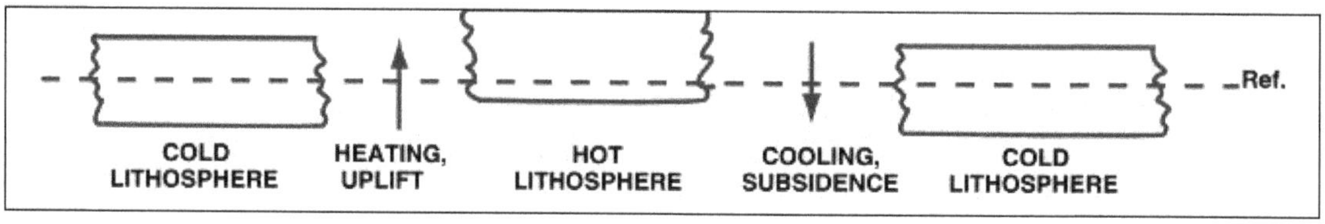

But suppose that some erosion took place while the crust was elevated. The crust is thinned where the erosion took place (and thickened somewhere else, where there was deposition; that might be far away, at the mouth of some long river system), so when the crust cools again it subsides to a position lower than where it started, thus creating a basin available for filling by sediments.

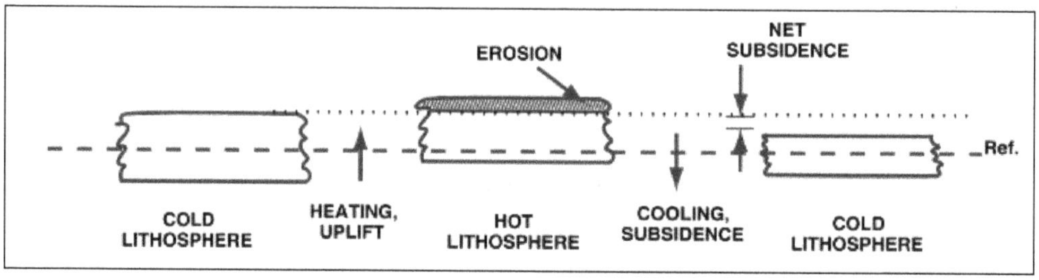

But the magnitude of crustal lowering by this mechanism is less than is often observed in basins thought to be created thermally. It has therefore been proposed, and widely accepted, that in many cases extensional thinning of the lithosphere accompanies the heating. Then, upon recooling, the elevation of the top of the lithosphere is less than before the heating and extension. This kind of subsidence has been invoked to explain many sedimentary basins.

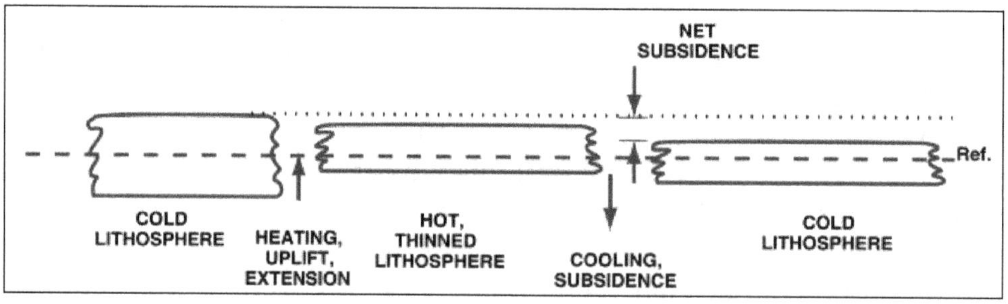

Another important way to make basins is to park a large load on some area of the lithosphere. The new load causes that lithosphere to subside by isostatic adjustment. But because the lithosphere has considerable flexural rigidity, adjacent lithosphere is bowed down also. The region between the high-standing load and the lithosphere in the far field (in the parlance of geophysics that just means far away) is thus depressed to form a basin. This model has been very successful in accounting for the features of foreland basins, which are formed ahead of large thrust sheets that move out from orogenic areas onto previously undeformed cratonal lithosphere.

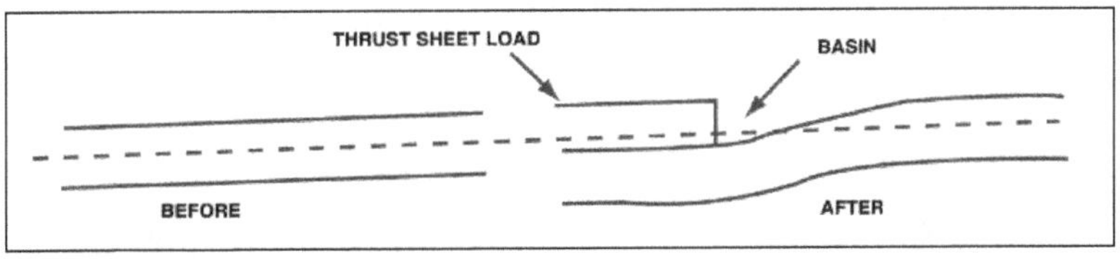

Geosynclines

The concept of geosynclines was developed in the last century to deal with the existence of thick successions of sedimentary rocks in what we would today call orogenic belts. A geosyncline is large troughlike or basinlike down-warping of the crust in which thick sedimentary and volcanic rocks accumulated. Usually, but not always, such accumulations are deformed during a later phase of the same geological cycle in which they were deposited. You can see by the definition that there is a close, although not one-to-one, correspondence between geosynclines and what we are discussing here as sedimentary basins.

The geosyncline concept was developed in an effort to understand the regularities of sedimentation in orogenic belts. Over the decades, in both Europe and North America, the concept was elaborated to an extreme degree, with lengthy classifications and polysyllabic terminology. The problem was that geologists were able to recognize distinctive kinds of sedimentary basin fills associated with orogenic belts, and characteristic histories of subsequent deformation of those sediment fills, but no one really knew the tectonic significance of geosynclines.

The universal acceptance of plate tectonics has provided a rational framework in which to interpret the development and history of sedimentary basins once called geosynclines. Plate tectonics has simply made the geosyncline concept obsolete.

About the only term in common use that is left over from the heyday of geosynclines is miogeocline, for the prograding wedge of mostly shallow-water sediment at a continental margin. The sediment thickens sharply oceanward and pass into thinner deep-water sediments.

Sedimentary Structure

Sedimentary structures are features that form in sediment as it is being deposited. These structures are typically an indication of what the sedimentary environment was like. When looking at structures in sedimentary rocks, they can often be identified by observable patterns in the sedimentary bedding or distinct shapes within the sediment. Basically, if the sedimentary rock doesn't look uniform and has a distinctive feature, there's a good chance it's a sedimentary structure.

Types of Sedimentary Structures

The simplest sedimentary structure is stratification, which is layering that can be observed in sedimentary rocks. Layers of sediment that are thicker than 1 cm are called beds and layers thinner than 1 cm are called laminations. Laminations are typically composed of fine-grained silt and clay-sized sediment. Structures can be more complex like the wavy pattern seen in ripple marks or chaotic looking patterns in cross-bedding.

Ripples caused by waves: This is a view from the top surface and not the side. This rock is Permian in age from Nomgon, Mongolia.

Cross-bedding in sandstone seen on a cliff face in Zion National Park, Utah. You can figure out the scale of the image by looking for trees or other vegetation.

Sedimentary structures provide a lot of information about the environment in which they formed including processes that were occurring when sediment was deposited, the environment of deposition, and the direction sediment was traveling, and/or the mechanism for transporting the sediment (wind, water, or ice). Some sedimentary structures also help you determine which side of the rock was originally facing upwards, called way-up indicators. When outcrops have overturned rocks (rocks that have been tilted so far they are upside down), sedimentary structures can be used to tell which way was originally facing up. Each structure tells a story that geologists use to interpret Earth's history.

Dunes and Ripple Marks

As water or wind moves across sediment, it can shape the grains into wavy patterns called dunes (>10 cm) and ripples (<10 cm). Symmetrical ripple marks are formed by the back-and-forth flow of water over sediment. These types of ripples are formed in the shallow marine environment where the back-and-forth motion of waves, or even tides, shape the sediment at the bottom of the ocean. These ripples have symmetrical limbs, meaning that both sides of the ripple dip at about the same angle. Symmetrical ripple marks are most commonly found in sandstones from shallow marine environments.

Examples of symmetrical ripple marks with limbs at the same angle on either side of the structure. A) Modern symmetrical ripples from the Bahamas. B, C, and D) Symmetrical ripples in Devonian-Missippian age sandstone.

Water moving in one direction, like a river, can produce asymmetrical ripple marks. The limbs on these ripples are not equal, with one side that is shallower and one side that is steeper. These types of ripple marks can tell you which direction the river was flowing because sediment moves up the shallow side of the ripple and gets deposited on the steep side. The deposition on the steep side of the ripple allows the ripple to move in the same direction that water is flowing. Wind can also create asymmetrical ripple marks at different scales. Ripple marks at smaller scales can usually be found along a beach. Large-scale ripple marks are called dunes, and are common in deserts and some coastal environments.

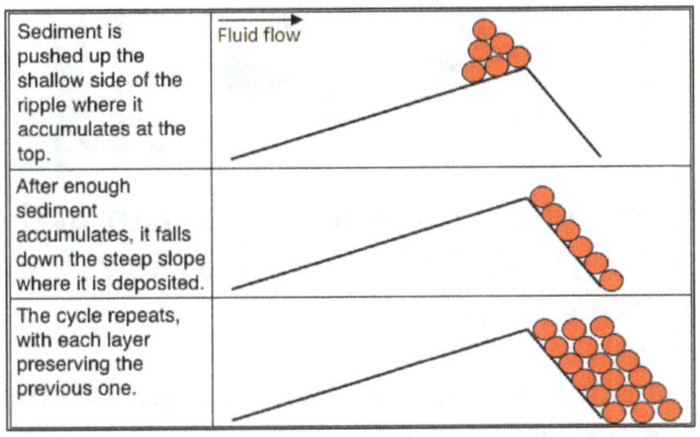

How asymmetrical ripples form: The red circles represent grains of sediment that move up a shallow side and fall down a steep side to form cross-bedding.

Examples of asymmetrical ripple marks. The limbs of the ripples dip at different angles; one shallow and one steep. Three views (A, C and D) are from the top and B is from the side. A) Modern asymmetrical ripples from the Bahamas. B and C) Asymmetrical ripples in sandstone from Colorado, USA. D) Precambrian asymmetrical ripples in quartzite.

Cross-Bedding

The top layer of a ripple or dune is not always preserved in the rock record, so it is rare to find ripples. Dunes and ripples are constantly moving, and as one passes and deposits its sediment, another follows right behind it to deposit more sediment on top. What geologists typically find in the rock record is the deposited sediment from the steep side of a series of ripples or dunes. The deposition of the steep side of several dunes or ripples creates a sedimentary structure called cross-bedding. One of the most important pieces of information that cross-bedding gives geologists is the direction that wind or water was moving. The steep side of a ripple always angles downward toward the direction the water or wind was moving. There are many different types of cross-bedding and each form in a similar way.

Cross-bedding from ancient sand dunes in Coyote Gluch, part of the Canyons of the Escalante, Utah. The upper image is uninterpreted, the lower image shows interpretations of four dunes as yellow dashed lines and the cross-beds in blue.

Sedimentary structures are not limited to Earth since similar features have been found on Mars, Venus, and Titan, Saturn's largest moon. Figure below shows cross-bedding from Mars, and it looks very similar to the wind-blown sand outcrops commonly found in the southwestern U.S. Do you think the scale is similar between these two images? The size of the cross-bedding can help to determine if these formed in water or air (aeolian). Smaller ripples form in water while larger ones form in terrestrial dunes.

An outcrop of cross-bedded sandstone on the lower slope of Mars' Mount Sharp. The direction of sediment transport is interpreted as sediments carried by currents moving down the deltas and into deeper lake water.

Graded Bedding

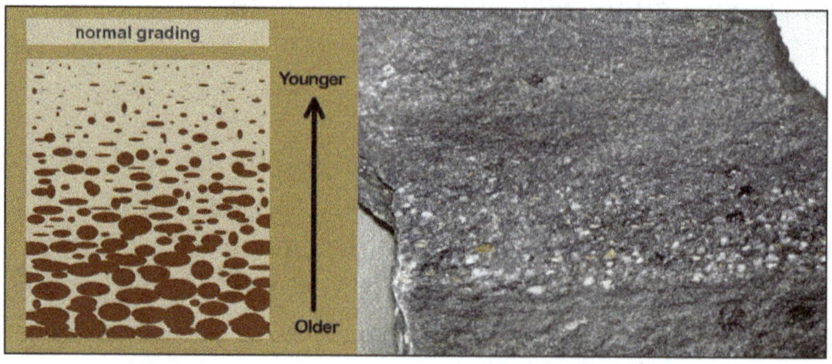

A sketch and an example of graded bedding. The left side of the figure is a sketch of graded bedding showing larger grains at the bottom and getting finer towards the top. The right side of the figure is a sample of graded bedding.

Graded bedding is a common sedimentary structure where a change in grain size can be observed within a single sedimentary bed. At the bottom of the bed are mainly coarse particles which get progressively smaller as you move vertically up the bed. Graded beds generally represent depositional environments in which transport energy decreases over time, like the changing water velocity in a river. However, these beds can also form during rapid depositional events, most commonly from turbidity currents. Turbidity currents are essentially underwater avalanches of sediment that move downslope, usually starting at the edge of the continental shelf and flowing down the continental slope. The sediment deposited from turbidity current is called a turbidite, which often has graded bedding with the coarsest particles at the bottom of the bed and the smallest at the top.

Mudcracks

Mudcracks, also called desiccation cracks, form when wet sediment, typically clay-rich, dries out. Clay minerals expand when they get wet and shrink when they dry out. As the sediment shrinks, cracks can develop, which form polygons on the surface of the mud. Today, you can find plenty of modern mudcracks along the margins of rivers or in desert valleys that periodically get inundated with floods. After a mudcrack forms, it can be filled in with new sediment.

Figure shows: (A) A modern example of large mudcracks in a dried up river bed in the Rio San Juan, Argentina. (B) An ancient example of mudcracks with sediment filling in the cracks from Maryland.

Mudcracks are typically wider at the top of the crack and get progressively smaller toward the bottom of the crack. Because of this pattern, mudcracks can be a good way-up indicator if you are able to see a cross-section view of the crack.

Mudcracks: The view spans about 4 feet (1.2 meters) left-to-right and combines three images taken by the Mars Hand Lens Imager (MAHLI) camera. The cracks may have formed more than 3 billion years ago.

Mars also has mudcracks, one of the pieces of evidence that indicates the red planet used to have liquid water on its surface. These were found in Gale crater in an exposure of Murray Formation mudstone on lower Mount Sharp. The white material in the cracks may be a form of calcium sulfate, either anhydrite or gypsum. This is a guess since Curiosity rover cannot test the mineral hardness.

Raindrop Impressions

Raindrop impressions are small, concave imprints made by rain when it falls on soft sediment. The impression or small crater is made from the force of the raindrop falling onto the sediment, which make these structures good way-up indicators. If you were to see only the bottom of the impression, it would look like a raised bump (convex).

Raindrop impressions tend to be found in fine-grained rocks like siltstones and shale but not in coarser-grained sandstones. The impressions likely represent the end of a rainstorm as rain is letting up because any previously formed impressions would be destroyed by subsequent rainfall. That's why most raindrop impressions are very scattered rather than occurring all over the surface. Then the impressions need to be filled in with sediment, before the next rainstorm, in order to be preserved.

A) Overhead view of modern raindrop impressions (and mudcracks) from Argentina. B and C) Ancient raindrop impressions from the Lower Permian, Upper Pecos Valley, New Mexico. B is an overhead view, C is the bottom-up view, showing the convex, underside of raindrop impressions (not the same sample as B).

Sole Marks

Sole marks is a broad term that describes a number of different sedimentary structures that appear as impressions or grooves in sediment, including flute casts, tool marks, groove casts, and load casts. Typically the cast of the marking (the raised bump) is preserved at the bottom of a sedimentary bed, hence the term "sole" mark, and the mold side (the impression) is filled with sediment. This makes sole marks good way-up indicators since the cast side is facing down.

Flute casts are common structures created by turbidity currents. The movement of these sediment avalanches underwater can scour the ocean floor, creating an elongated impression. Flute casts are usually closely spaced and can be stacked on top of one another. Not only can they tell you which way is up, but they can also tell you which way the current was flowing. The tapered end of the flute cast points in the direction of flow.

Flute casts from the central Alps, Switzerland. The view is from the underside of the rock.

Tool marks are made when an object, such as a stick, is dragged across sediment by a current and leaves behind what looks like scratches in the soft sediment. The elongated scratches can be used as an indicator of the paleocurrent.

Tool marks from Banff National Park. The view is from the underside of the rock.

Groove casts are raised parallel ridges. They are spaced closely together, often appearing in sets of 2 and 3, but do not occur on top of one another like flute casts. Interpreting the paleocurrent from groove casts can be difficult because the marking is often symmetrical. Without the addition of other paleocurrent evidence, you may only be able to narrow down the paleocurrent to two directions that are 180° apart.

Groove casts: The yellow arrow indicates the direction of transport. This arrow is double ended as there is no indication if the water was flowing up or down the river before it was tilted and exposed. The view is from the underside of the rock.

Load casts form when dense, sandy sediment is deposited on less dense, water-saturated sediment, usually silt or clay. The dense sand load pushes into the soft layer below, creating bulb-like impressions.

Load casts in arkose sandstone from the Aquitaine Basin near Nontron, France. The view is from the underside of the rock.

Sedimentary Structures and Paleoenvironments

Sedimentary structures are incredibly useful for determining what paleoenvironments were like. By combining sedimentary structures and the surrounding geology, a geologist could describe a pretty accurate picture of what the environment was like when these sediments were being deposited.

Trace Fossils and Bioturbation

Organisms borrow and move through sediment on the ocean floor and the bottoms of lakes and rivers; this is called bioturbation. When organisms disturb the sediment by borrowing, their borrows can be preserved when the sediment hardens into rock. In most cases, the borrows will fill with new sediment, but the outline is preserved. It is difficult to assign a specific organism to the creation of a single borrow. Instead, geologists look at different borrows that tend to occur together in the rock record in order to classify them, a branch of study known as ichnofacies. Each ichnofacies is named after the most common trace fossil in the facies. Determining which ichnofacies the trace fossil fits into can tell you about the environment in which the organism lived, including water depth, salinity, energy and turbidity, and what the substrate was like. Generally, vertical borrows were formed in shallow water environments while horizontal burrows in deeper water environments.

Table: Common ichnofacies.

Ichnofacies	Substrate	Paleoenvironment	Description
Scoyenia	Sandstones may be associated with red beds.	Terrestrial, freshwater, low energy.	Horizontal, curved, and rope-like burrows. Unbranching, but can cross each other. Occasional vertical burrows.
Psilonichnus	Variable grain size, sand.	Coastal, barrier islands, deltas, estuaries, lagoons, bays.	Vertical burrows with J, Y, or U shapes. Can also have vertebrate footprints.
Trypanites	Hard rock, carbonate, shells.	Coastal cliffs, reefs, beachrock.	Closely-spaced straight, or slightly curved, vertical burrows.
Glossifungites	Firm but not lithified sediment.	Shallow marine, marginal marine, delta, estuary.	Three-dimensional network of cylindrical burrows and individual, vertical, teardrop-shaped burrows.

Skolithos	Sand.	Beaches, tidal flats, shallow marine, above wave base.	Straight, vertical, burrows that do not branch or cross. Can be slightly angled and J-shaped.
Cruziana	Sand and silt.	Mid to distal continental shelf, below wave base, but above storm wave base.	Horizontal and vertical burrows from a wide variety of organisms.
Zoophycos	Sand and silt.	Deep water, base of continental shelf, may be associated with turbidites.	A series of horizontal burrows curving away from a central point.
Nereites	Silt and clay.	Deep water, open ocean.	Meandering and spiraling horizontal trails or burrows.

Sedimentary Process

Sedimentary processes, namely weathering, erosion, crystallization, deposition, and lithification, create the sedimentary family of rocks. Most of Earth's surface is covered by sediments (like mud, sand, and gravel), and the majority of rocks you will see (the majority of rocks exposed at Earth's surface) are sedimentary rocks (like mudstone, sandstone, and limestone).

Chemical Weathering and Mineral Resources

Weathering processes break rocks into smaller pieces. There are two types of weathering: chemical and mechanical. Chemical weathering changes the chemical make-up of minerals and hence breaks them apart. For example, when sugar is stirred into water, the sugar dissolves (the sugar crystal breaks apart into the molecules that compose it; it is no longer a crystalline solid). This is a type of chemical weathering. Rust is also an example of chemical weathering; in that case, iron minerals oxidize and become smaller minerals with different compositions. There are several things to keep in mind about chemical weathering:

- Chemical weathering requires liquid water: There will be more chemical weathering where rainfall and humidity are high.
- Different minerals chemically weather and dissolve differently (they have different solubilities). Some minerals such as halite (salt) dissolve easily (quickly). Some minerals will not dissolve. Some minerals might not dissolve quickly, but they may succumb slowly or to other types of chemical weathering. Stable minerals do not chemically weather very easily or at all.
- Chemical weathering can only act on the surface of a rock or mineral, or where a crack allows air and water into the mineral/rock. Therefore, more chemical weathering happens around these edges where the rocks have first been mechanically weathered. Mechanical weathering physically breaks a rock into smaller pieces of the same rock or minerals in that rock.

Mineral Solubility and How Chemical Weathering Creates Mineral Reserves

Table: Mineral Solubilities.

Mineral	Time it would take a hypothetical 1 mm sphere of this mineral to dissolve in water with ph=5 (years)
Calcite	0.1
Olivine (forsterite)	2300
Plagioclase (albite)	575,000
Mica (muscovite)	2,600,000
Clay (kaolinite)	6,000,000
Quartz	34,000,000

The above table lists how long it would take for very small pieces of minerals to dissolve. This list does not include halite (salt), which would dissolve in minutes instead of months or years. The minerals are listed from most soluble to least soluble. Halite and calcite have high solubilities, and quartz has a very low solubility and is thus more stable.

Not only do minerals have different solubilities, but elements within minerals dissolve differently. When minerals chemically weather, some elements in those minerals will readily dissolve in water and be removed by water. Other elements are not readily soluble, so will remain in a mineral. The mineral may not completely break into separate ions but will change composition (and hence become a different mineral) as one type of ion is removed.

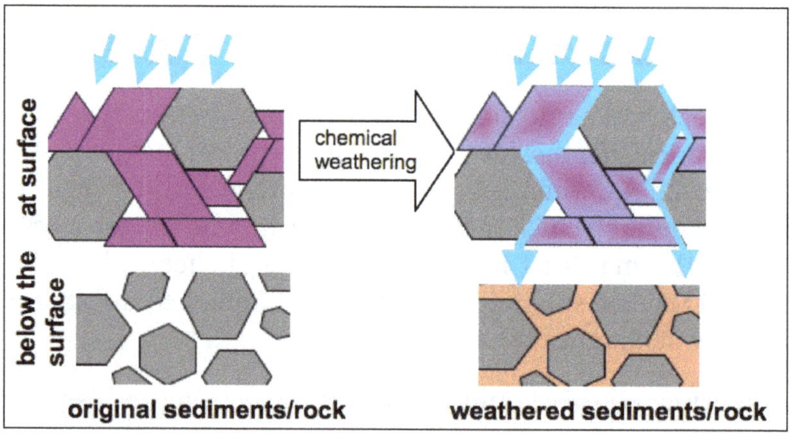

The colored shapes represent minerals, and white space is pore space (space in rock that is filled with air or water). The original sediment (or rock, on left) at the surface is exposed to rain (the arrows from above) and hence, chemical weathering. The gray minerals are not soluble, so remain unchanged. The purple minerals are made of red and blue elements. The purple minerals react with the water, which preferentially removes one element (the red) from the mineral and carries that element deeper into the ground, where it crystallizes as a new mineral. Weathering therefore separated the blue from the red elements, and concentrated those elements. A mining company interested in the blue element will mine rocks/sediments right at the surface, whereas a mining company interested in the red element will extract deeper rocks/sediments.

For example, imagine that a rock is made of minerals X and Y (purple and gray in the figure to the right, respectively). Mineral Y is made of elements C, D, and E. Mineral X is very stable, and does not react with water, similar to quartz. Water flowing through the broken rock will alter mineral Y, but X will stay intact. In mineral Y, imagine that element C (the red element in the figure) is soluble, but elements D and E are not. Water will remove C. The concentrations of elements left behind (D and E: the insoluble elements) increase in the mineral; chemical weathering thus concentrated D and E (D and E would be found in residual weathering deposits also called laterite deposits). In addition, the water containing the dissolved ions will carry element C somewhere else, where C may join with other ions and crystallize to make a mineral; and in this way, C is also concentrated. Thus chemical weathering concentrates minerals and/or elements, and turns mineral resources into mineral reserves.

Because they form right at the surface, residual weathering deposits are likely to erode away. Therefore, residual weathering deposits tend to be found in fairly young sediment layers, as older deposits will have already eroded.

The characteristic red color of the Waimea Canyon on the island of Kauai, Hawaii, results from chemical weathering. The original lava flows were chemically weathered by rainwater and the Waimea River, dissolving out the soluble elements, and concentrating relatively insoluble, rust-red iron oxide (hematite) on the surface. The red mud of the Waimea Canyon is used today as a resource by a local company to color their t-shirts.

Table: Mineral Resources Formed by Chemical Weathering.

Mineral commodity	Commodity	How the mineral forms	Some uses
Gibbsite, boehmite (in the rock called bauxite)	Aluminum	Chemical weathering of aluminum-rich sediments, igneous rocks, or volcanic ash.	Food and beverage packaging, parts in airplanes and cars.
Garnierite and goethite	Nickel	Chemical weathering of ultramafic (low silica content) igneous rocks.	Stainless steel, batteries.
Goethite and hematite	Iron oxides, iron	Not currently mined for iron, but is a potential reserve.	Steel, pigment (paint, makeup).
Clay minerals	Clay	Chemical weathering of feldspar minerals.	Ceramics, paper, cat litter.

Mechanical Weathering, Erosion, Deposition and Mineral Resources

Erosion and Deposition

Mechanical weathering processes break rocks into smaller pieces without changing their compositions. The rock and mineral fragments created by mechanical weathering are called clastic sediments, and will be eroded (picked up and carried) and deposited elsewhere. Thousands or millions of years of erosion can deposit large quantities of a certain mineral in one spot, so even though the concentration of the mineral might have been low in the weathered rock, the concentration can be much higher in the sedimentary deposit. These mineral deposits are called placer deposits.

Streams are often the agents that erode (transport) these minerals. Streams can deposit sediments along the stream bank (in sand or gravel bars) and in alluvial fan or delta deposits. Once the sediments deposit in the ocean, they can be eroded further by waves and deposited in beaches. Wind can also pick up clastic sediment and concentrate placer deposits.

The sand on this Lake Superior beach is clastic sediment. The sand was created when a rock mechanically weathered (broke into pieces), and then was eroded (carried) by streams to Lake Superior, where it deposited. Further erosion by the waves concentrates the sand on the beach.

Sediments deposited by wind, streams, and waves are sorted by size and density. Gravity pulls sediments to the ground (or the bottom of the stream or ocean), but the wind or rushing water (or waves) pulls sediments sideways, attempting to keep them aloft. Larger and denser (higher specific gravity) sediments require more wind, stream, or wave energy to stay afloat.

If we know the size and specific gravity of the mineral that we are looking for, and we know the energy levels of a stream (or wave), then we can predict where in the stream (or shoreline) we should look for that mineral.

For example, gold tends to form small clastic sediments, but it is very dense. Therefore, small pieces of gold may deposit with large pieces of less dense clastic sediment (gravel) along streams. Therefore we should look for gold in gravel bars in modern streams, or in deposits of conglomerate that were once gravel bars.

Before you go to the local stream to pan for gold, remember that the rock that weathered to begin with is important. If the rock that mechanically weathered to make the gravel did not contain any gold, then we will not find gold in the gravel. Gold is mostly found in igneous or metamorphic rocks. So, a gravel bar along the Meramec River in Missouri that contains only the chemical sedimentary rock chert is unlikely to contain gold. A river flowing from a granite outcrop in the Rocky Mountains, however, might have gold in its gravel bars.

This lake in Edwardsville, Illinois fills an abandoned shale quarry. This shale is a clastic sedimentary rock formed when mud deposited in a swampy area about 300 million years ago and since lithified. Before reclamation laws were in place, the shale was mined to make bricks. The quarry was not filled in and returned to its pre-mining state, but left as a lake.

Sand and gravel are important industrial mineral resources. Sand and gravel are clastic sediments, so may be eroded and deposited in environments (e.g., along streams and beaches) that can later be mined for these resources.

Table: Mineral Resources Formed by Mechanical Weathering, Erosion and Deposition.

Mineral	Commodity	Specific gravity	Some uses
Various	Gravel		Construction.
Quartz	Sand, silica	2.65 (quartz)	Construction, glass, computer chips.
Rutile, ilmenite	Titanium	4.18-4.25 (rutile), 4.7 (ilmenite)	White pigment (paint), titanium metal (airplanes).
Chromite	Chromium	4.6	Stainless steel.
Rare-earth oxides	Rare-earth elements (REE)		Computer and television displays, fiber optic cables, magnets, catalytic converters in cars.
Cassiterite	Tin	6.8-7.1	Coating for tin cans, bronze.
Cinnabar	Mercury	8.1	Fluorescent lights, thermometers, production of chlorine.
Diamond	Diamond	3.5	Gemstone, abrasive.
Native copper	Copper	8.9	Wiring, antifungal and antibacterial chemicals.
Native silver	Silver	10.5	Mirrors, photography, electronics.
Native gold	Gold	15-19.3	Jewelry, electronics, currency.
Native platinum	Platinum	14-19	Jewelry, catalytic converters in cars, dental equipment.

Chemical Sedimentary Rocks and Mineral Resources

Water contains the dissolved ions derived from chemical weathering. When water evaporates, only pure water will turn to vapor and enter the atmosphere. Any ions are left behind, and the concentration of ions in the remaining water increases. When the water becomes saturated in ions, those ions will combine to form new minerals (i.e., crystallization will occur). The new mixture of minerals is a chemical sedimentary rock. Chemical sedimentary rocks that form when water evaporates are called evaporite deposits.

Table: Compositions of Brines.

Ion	Average seawater (mg/kg)	Great Salt Lake (mg/kg)	Rain water from Menlo Park, CA (mg/kg)	Mississippi River (mg/kg)
Chloride	19,350	140,000	0.8-17	24
Sodium	10,760	85,600	0-9.4	20
Sulfate	2,710	16,400	0.7-7.6	51
Magnesium	1,290	7,200	0.7-1.2	10
Calcium	411	241	0.8-1.2	38
Potassium	399	4,070	0	2.9
Bicarbonate	142	251	7-Apr	113
Silica	.5-10	48	0.3-1.2	7.9

Composition of average seawater and brine from Great Salt Lake, Utah and freshwater (rain and river water). The units are mg/kg, which gives the mass (in milligrams) of an ion (like sodium) in one kilogram of water. In the ocean and the Great Salt Lake, evaporation removes water, leaving other ions behind. This concentrates the ions (as you can see by comparing the amounts of ions in seawater and the Great Salt Lake with the ions in rain water and the Mississippi River). Brines, like that in the Great Salt Lake, are saltier than seawater; and while all brines are saltier than seawater, different brines will contain different amounts of certain ions.

As water evaporates, different ions reach saturation, and thus crystallize, at different times. The most insoluble minerals crystallize first, and the most soluble crystallize last. So, as salt water evaporates, calcite will crystallize first (using up the carbonate and bicarbonate ions), then gypsum (using up the sulfate), then halite (using up the sodium), and lastly potassium and magnesium salts. The minerals (and rocks they comprise) form in distinct layers, with calcite (which forms the rock limestone) on the bottom and potassium and magnesium salts on top. Because these chemical sedimentary rock contain only one mineral, beneficiation of the mined reserve (where minerals have to be separated from other minerals) may not be needed.

A limestone quarry in Dupo, Illinois: Illinois is nowhere near the ocean today, but was covered by a shallow sea when this limestone formed about 300 million years ago. The limestone is mined to make cement.

Because high evaporation rates are needed for crystallization, climate is important. For example, limestone (the sedimentary rock made of calcite) is presently forming in the shallow seas off the Bahamas. Also, natural highs and lows in the seafloor and/or coral reefs can create basins that favor chemical sedimentary rock formation.

Table: Chemical Sedimentary Rocks that are Mineral Resources.

Rock	Mineral	Commodity	Some uses
Limestone, dolostone	Calcite, dolomite	Limestone, calcite (calcium carbonate)	Cement, calcium in food and supplements
Rock gypsum	Gypsum	Gypsum	Drywall
Rock salt	Halite	Halite (salt), chlorine (element)	Food additive, road deicer, source of chlorine (water treatment, bleach, etc.)
Rock salt	Sylvite	Potassium salt	Food additive, road deicer, source of potassium (fertilizer)
Banded iron formation	Magnetite, hematite	Iron	Steel

One nice thing about evaporite minerals is that they are readily soluble; that is, it is easy to remove the desired element from the mineral. For example, if soybean plants need sulfur, ground gypsum (a mineral) can be added to the soil, and the soybeans can readily remove the sulfur from the gypsum. Humans need sodium. If we eat the mineral halite, then our bodies easily remove the sodium from the salt. It would be almost impossible to remove sodium from an insoluble silicate mineral like albite (feldspar), for example:

Instead of mining the crystallized evaporite deposits, salt water or brines (even saltier water) can be mined. Salt water is, of course, found in the oceans. Brines are found in terminal lakes (lakes into which rivers flow but which have no outlets) as well as underground (the deepest groundwater is often brine).

Sedimentary Rocks and Processes

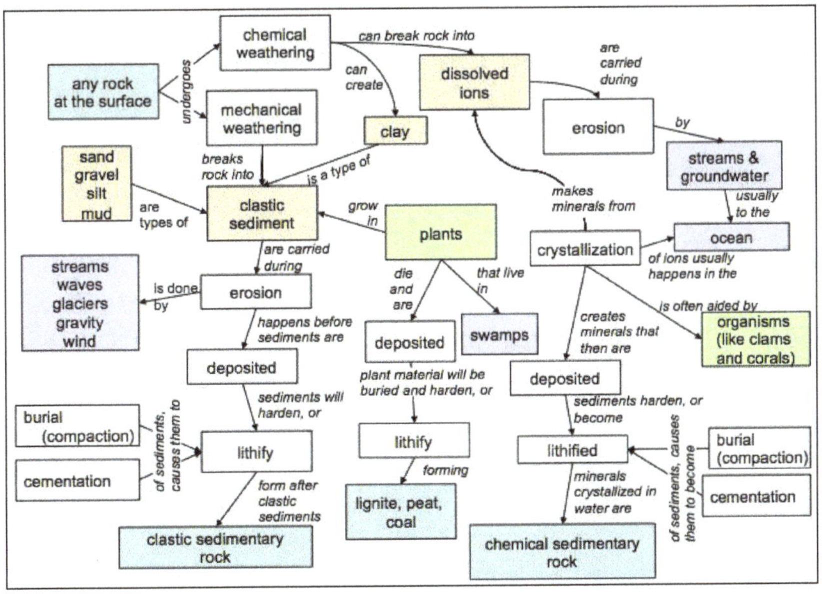

Concept map showing sedimentary processes and sedimentary rocks.

The concept map shows how all sedimentary rocks form. The blue and tan boxes contain substances, the purple boxes show places or forces, and white boxes contain sedimentary processes (the actual events that create sedimentary rocks). These processes are further defined in the glossary. Some sedimentary rocks form with the help of living things (green boxes). (Rocks that form from or with the help of organisms are sometimes called biochemical or bioclastic sedimentary rocks). The different sedimentary processes can concentrate elements, making them economical to mine. In other words, sedimentary processes can turn a mineral resource into a mineral reserve.

References

- Sedimentary-rock, science: britannica.com, Retrieved 07, March 2020
- Sediment-gravity-flow-the-conceptual-problem-222022131: researchgate.net, Retrieved 11, August 2020
- Sediment-budget, earth-and-planetary-sciences: sciencedirect.com, Retrieved 04, July 2020
- Sedimentary-geology-spring, earth-atmospheric-and-planetary-sciences: ocw.mit.edu, Retrieved 19, May 2020
- Sedimentary-structures, historicalgeologylab: uhlibraries.pressbooks.pub, Retrieved 16, April 2020
- Mineral_resources: serc.carleton.edu, Retrieved 21, March 2020

Chapter 6

Integrating Meteorology and Climatology

Meteorology is the study of the atmosphere and the various phenomena associated with it along with their effects on the weather. Climate and global weather conditions are influenced by the atmosphere and its study is known as climatology. This chapter closely examines both these area of study in order to provide a thorough understanding of the related topics.

Meteorology

Meteorology is the scientific study of the atmosphere that focuses on weather processes and forecasting.

Meteorological phenomena are observable weather events which illuminate and are explained by the science of meteorology.

Those events are bound by the variables that exist in Earth's atmosphere.

They are temperature, pressure, water vapor, and the gradients and interactions of each variable, and how they change in time.

The majority of Earth's observed weather is located in the troposphere.

Although meteorologists now rely heavily on computer models (numerical weather prediction), it is still relatively common to use techniques and conceptual models that were developed before computers were powerful enough to make predictions accurately or efficiently.

Types of Meteorology

Boundary layer meteorology studies the layer of air that is directly above the Earth's surface. It focuses on how the surface-atmosphere boundary affects the oceans, lakes, urban land areas and non-urban land areas. This study delves into how the heating, cooling, and friction affects these areas and how these cause turbulent mixing in the boundary layer of air.

Mesoscale meteorology studies several layers of the Earth including the boundary layer, the stratosphere, the tropopause and the troposphere. It examines several forms of weather such as land breezes, mountain waves, thunderstorms, fronts, squall lines and precipitation bands in extratropical and tropical cyclones. These weather events are then often followed throughout their duration.

Global scale meteorology studies the weather patterns that are significant to the transport of heat from the north and south poles to the tropics. It also focuses on oscillations of a vast scale. It delves into how global oscillations cause weather disturbances and climate disturbances in mesoscale timescales and synoptic timescales.

Agricultural meteorology studies how weather affects plant development, animal development, plant distribution, crop yield and the efficiency of water use. This type of meteorology studies the energy balance of natural ecosystems and managed ecosystems for similarities and differences. Those in this field are also looking into how vegetation may affect the weather and climate.

Hydrometeorology studies the hydrologic cycle tracks the storms' rainfall statistics and deals with the water budget. This type of meteorologist prepares the forecasts that are often in the news and gives the information on amounts of potential precipitation, snow, and rain and predicts areas that may be in danger of flash floods.

Measured Parameters of Meteorology

Meteorological measurements for weather forecasting and climatology have been carried out on a regular basis for centuries. However, the data acquired can only be evaluated and interpreted after having statistically recorded medium-term and long-term atmospheric conditions.

Nowadays, transport and communications such as ground, air and sea traffic could not be maintained without these data, which are mainly being collected from measurements and observations in the atmosphere close to the ground (the Boundary Layer). The main meteorological parameters in this field are:

- Wind speed and direction.
- Air temperature.
- Air pressure.
- Air humidity.
- Precipitation.
- Haze and contents of the air.
- Solar and terrestrial radiation.

These parameters are also significant for such issues as air pollution, avalanche warning, sun simulation, renewable energy industry, agriculture, forestry, water supply and distribution, town and regional planning. For example, the evaluation and interpretation of gas emission measurements is only possible in comparison with meteorological data acquired concurrently.

The structure of the atmosphere close to the ground is extremely important for the local climate. Knowing the solar radiation as well as the air humidity and air temperature is necessary to evaluate chemical reactions of pollutants in the air.

All meteorological parameters are subject to short-term variations, normally caused by turbulences within the atmosphere. All meteorological parameters are influenced by solar radiation, directly or indirectly, and this results in typical daily or yearly trends.

In order to be able to evaluate these typical trends it is necessary to compute the mean values from the actual ones measured over a specific period. For some meteorological parameters it is quite easy to understand their daily cycle. For example, the temperature cycle is normally a simple curve with a minimum value shortly after sunrise and a maximum in the early afternoon. The yearly cycle of a meteorological parameter can be determined by making daily measurements. The average yearly cycle within a climatic region is normally determined by making measurements over a minimum of 30 years.

Of course, meteorological measurements have to be made outside. This means that the sensors and associated electronics must be designed to withstand the local climatic conditions, which may be extreme, from deserts to Antarctica.

Within the atmosphere close to the ground the temporal and spatial characteristics of radiation values are influenced by the characteristics of the ground surface. The most influential factors affecting the received radiation at any particular location are:

- Location on the earth.
- Date and time.
- Precipitation (cloud, fog, rain, snow).
- Constriction of the horizon (field of view).
- Air pollution (aerosols and gasses).
- Albedo.

Due to the physical effects mentioned above it is sometimes not sufficient for many application fields to just measure the 'Global Radiation' coming from all around the measurement location. It may also be necessary to measure the 'Direct Radiation' coming only from the sun and/or the 'Diffuse Radiation' (not coming directly from the sun). The 'Radiation Balance' of incoming to outgoing radiation in the short-wave and long-wave may also be required.

Before reaching the ground the solar radiation is influenced by our atmosphere and its physical characteristics and an essential parameter is the absorption in different wavelength ranges. Albedo is affected by surfaces with different reflection characteristics, such as water, ice, snow, stone, grass, crops or woodland.

The different wavelength ranges as well as the properties of the atmosphere and the ground surface must be taken into consideration and this makes it necessary to develop special sensors suitable for individual, very complex, measuring tasks.

In order to be able to develop the proper sensor, it is necessary beforehand to determine which meteorological value shall be measured and how it is defined.

The parameters within the short-wave spectral range are:

Direct Solar Radiation	S
Diffuse Sky Radiation	H
Global Radiation	G (= S + H)

Reflected Global Radiation	R
Albedo (Reflection Factor)	R / (S + H)
Short-wave Radiation Balance	(S + H) - R

The parameters within the long-wave spectral range are:

Atmospheric Radiation	A
Emission of Ground Surface (including reflected atmospheric radiation)	E
Downward Global Radiation	S + H + A
Upward Global Radiation	R + E
Long-wave Radiation Balance	A - E

Consequently, the radiation balance over the complete spectrum is the difference between received and returned radiation:

Radiation Balance	Q (= S + H – R + A - E)

These parameters represent the most important climatological factors, for which the following sensors have been developed.

Direct Solar Radiation

Sensor	Pyrheliometer
Spectral Range	300nm - 3,000nm
View Angle	< 5°
Neccessary Accessory	Sun Tracker

Global Radiation

Sensor	Pyranometer
Spectral Range	300nm – 3,000nm
View Angle	180°
Necessary Accessory	None

Reflected Global Radiation

Sensor	Pyranometer
Spectral Range	300nm – 3,000nm
View Angle	180°
Necessary Accessory	Glare Shield to keep out low-angle non-reflected radiation

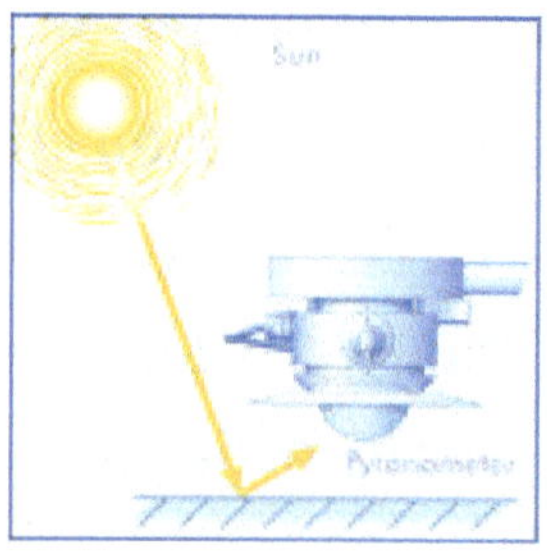

Diffuse Sky Radiation		
Sensor	Pyranometer	
Spectral Range	300nm – 3,000nm	
View Angle	180°	
Necessary Accessory:	Shadow Ring	
Shading Device:	Shadow Band	
	SunTracker/Shadow Ball	
	Sun Tracker/Shadow Disk	

Atmospheric Radiation		
Sensors	Pyranometer and Pyrgeometer	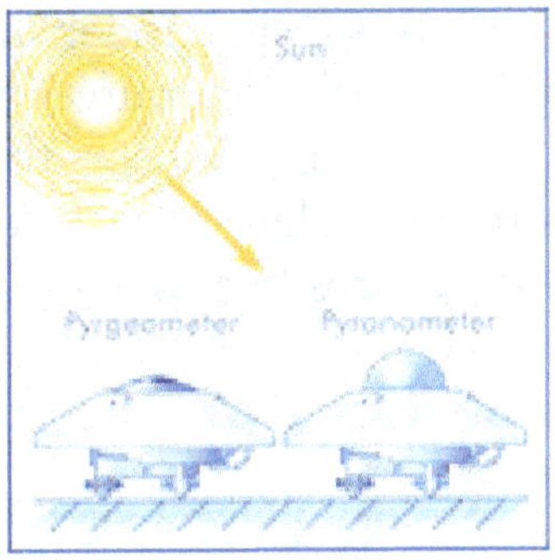
Spectral Range	300nm – 50µm	
View Angle	180°	
Necessary Accessory	None	

Emission of Ground Surface		
Sensor	Pyranometer and Pyrgeometer	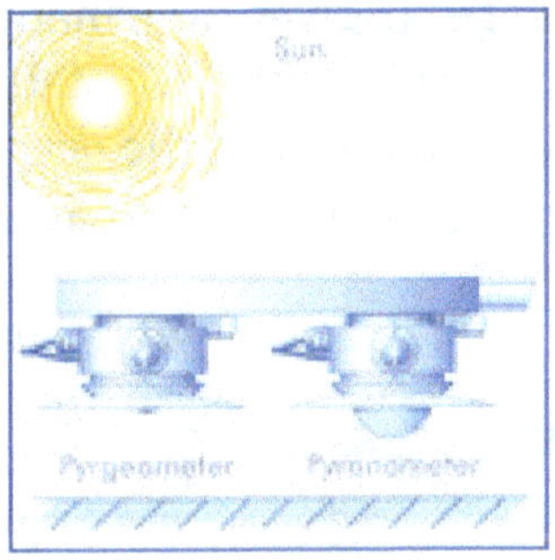
Spectral Range	300nm – 50µm	
View Angle	180°	
Necessary Accessory	Glare Shields to keep out low-angle non-reflected radiation	

Radiation Balance		
Sensor	Net Radiometer 2 Pyranometers and 2 Pyrgeometers	
Spectral Range	300nm – 50µm	
View Angle	180°	
Necessary Accessory	Glare Shields for downward facing sensors to keep out low-angle non-reflected radiation	

Combinations of these meteorological measurements are necessary for weather forecasts, climatology and other applications mentioned before, depending on the task and the accuracy required.

In order to monitor and minimise air pollution, measurements complying with the requirements of national standards for the evaluation of emissions and air quality are also made. These are termed 'environmental measurements'.

Usually these environmental measurements comprise the measurement of the global radiation, the direct radiation and the radiation balance. The data acquired from global radiation and direct radiation, together with specific spectral data help to determine the amount of gasses and aerosol particles in the air and also provide information on the photo-chemical formation of secondary impurities. Measuring the radiation balance provides information on the vertical exchange and spread of the pollution.

A further field of application for solar sensors is environmental simulation where the effect on materials is tested under 'artificial suns'. In order to cut down the testing time, very high intensity artificial radiation sources are used that far exceed the maximum natural value of 1,367 W/m^2 (the Solar Constant). Sensors that are used for these purposes should have measuring ranges that allow for radiation levels in the range 2,000 to 4,000 W/m^2 and they are often required to work at high temperatures (> 100°C).

Importance of Meteorology

Meteorology plays a significant role in environmental science. It is helpful for determining and tracking climate patterns as well as how land and water play a part in the climate and climate change. It gives information on oscillations and how global oscillations may cause weather and climate disturbances. Meteorology also helps to control vegetation, and it also allows those in agriculture determine the best time to promote their crops. Meteorology also studies how past climates can predict future climates and the dangers that certain environmental processes have on Earth.

Scales of Meteorology

Meteorological phenomena occur over a wide range of scales. The four meteorological scales are: microscale, mesoscale, synoptic scale, and global scale.

Microscale Meteorology

Microscale meteorology focuses on phenomena that range in size from a few centimeters to a few kilometers, and that have short life spans (less than a day). These phenomena affect very small geographic areas, and the temperatures and terrains of those areas.

Microscale meteorologists often study the processes that occur between soil, vegetation, and surface water near ground level. They measure the transfer of heat, gas, and liquid between these surfaces. Microscale meteorology often involves the study of chemistry.

Tracking air pollutants is an example of microscale meteorology.

Mesoscale Meteorology

Mesoscale phenomena range in size from a few kilometers to roughly 1,000 kilometers (620 miles).

Two important phenomena are mesoscale convective complexes (MCC) and mesoscale convective systems (MCS). Both are caused by convection, an important meteorological principle.

Convection is a process of circulation. Warmer, less-dense fluid rises, and colder, denser fluid sinks. The fluid that most meteorologists study is air. (Any substance that flows is considered a fluid.) Convection results in a transfer of energy, heat, and moisture—the basic building blocks of weather.

In both an MCC and MCS, a large area of air and moisture is warmed during the middle of the day—when the sun angle is at its highest. As this warm air mass rises into the colder atmosphere, it condenses into clouds, turning water vapor into precipitation.

An MCC is a single system of clouds that can reach the size of the state of Ohio and produce heavy rainfall and flooding. An MCS is a smaller cluster of thunderstorms that lasts for several hours. Both react to unique transfers of energy, heat, and moisture caused by convection.

Orlanski divides the "mesoscale" into three "submesoscales":

1. Meso- γ : 2-20 km.
2. Meso-β : 20-200 km.
3. Meso- α : 200-200 km.

Synoptic Scale Meteorology

Synoptic-scale phenomena cover an area of several hundred or even thousands of kilometers. High- and low-pressure systems seen on local weather forecasts, are synoptic in scale. Pressure, much like convection, is an important meteorological principle that is at the root of large-scale weather systems as diverse as hurricanes and bitter cold outbreaks.

Low-pressure systems occur where the atmospheric pressure at the surface of the Earth is less than its surrounding environment. Wind and moisture from areas with higher pressure seek low-pressure systems. This movement, in conjunction with the Coriolis force and friction, causes the system to rotate counter-clockwise in the Northern Hemisphere and clockwise in the Southern Hemisphere, creating a cyclone. Cyclones have a tendency for upward vertical motion. This allows moist air from the surrounding area to rise, expand and condense into water vapor, forming clouds. This movement of moisture and air causes the majority of our weather events.

Hurricanes are a result of low-pressure systems (cyclones) developing over tropical waters in the Western Hemisphere. The system sucks up massive amounts of warm moisture from the sea, causing convection to take place, which in turn causes wind speeds to increase and pressure to fall. When these winds reach speeds over 119 kilometers per hour (74 miles per hour), the cyclone is classified as a hurricane.

High-pressure systems occur where the atmospheric pressure at the surface of the Earth is greater than its surrounding environment. This pressure has a tendency for downward vertical motion, allowing for dry air and clear skies.

Extremely cold temperatures are a result of high-pressure systems that develop over the Arctic and move over the Northern Hemisphere. Arctic air is very cold because it develops over ice and snow-covered ground. This cold air is so dense that it pushes against Earth's surface with extreme pressure, preventing any moisture or heat from staying within the system.

Surface Weather Analysis

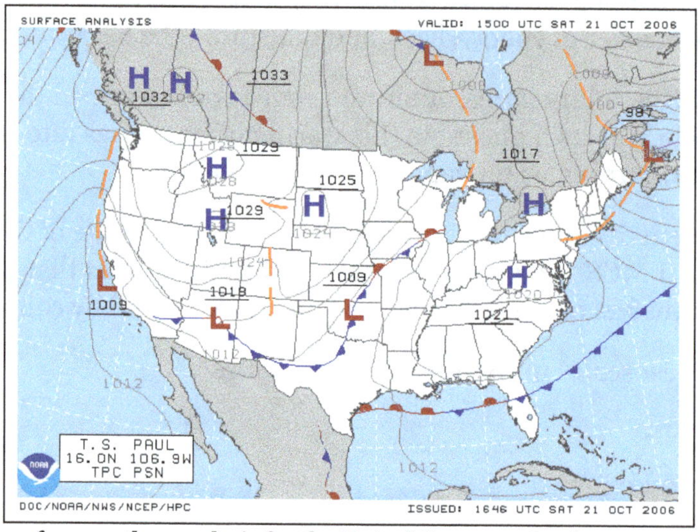

A surface weather analysis for the United States on October 21, 2006.

A surface weather analysis is a special type of weather map that provides a view of weather elements over a geographical area at a specified time based on information from ground-based weather stations. Weather maps are created by plotting or tracing the values of relevant quantities such as sea level pressure, temperature, and cloud cover onto a geographical map to help find synoptic scale features such as weather fronts.

The first weather maps in the 19th century were drawn well after the fact to help devise a theory on storm systems. After the advent of the telegraph, simultaneous surface weather observations became possible for the first time, and beginning in the late 1840s, the Smithsonian Institution became the first organization to draw real-time surface analyses. Use of surface analyses began first in the United States, spreading worldwide during the 1870s. Use of the Norwegian cyclone model for frontal analysis began in the late 1910s across Europe, with its use finally spreading to the United States during World War II.

Surface weather analyses have special symbols which show frontal systems, cloud cover, precipitation, or other important information. For example, an *H* represents high pressure, implying good and fair weather. An *L* represents low pressure, which frequently accompanies precipitation. Various symbols are used not just for frontal zones and other surface boundaries on weather maps, but also to depict the present weather at various locations on the weather map. Areas of precipitation help determine the frontal type and location. Mesoscale systems and boundaries such as tropical cyclones, outflow boundaries and squall lines also are analyzed on surface weather analyses. Isobars are commonly used to place surface boundaries from the horse latitudes poleward, while streamline analyses are used in the tropics.

Extratropical Cyclone

An extratropical cyclone is a synoptic scale low-pressure weather system that has neither tropical nor polar characteristics, being connected with fronts and horizontal gradients in temperature and dew point otherwise known as "baroclinic zones".

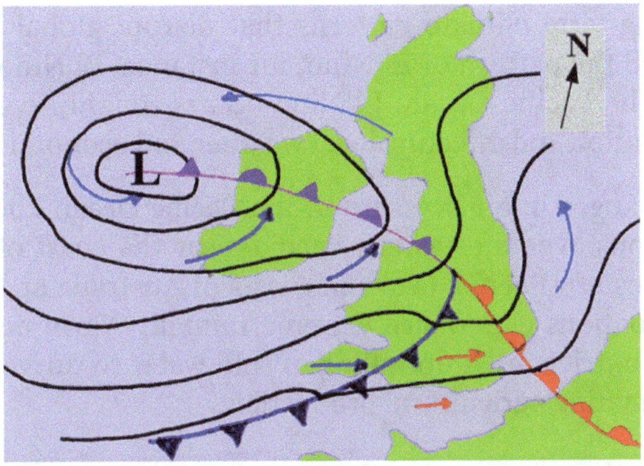

A fictitious synoptic chart of an extratropical cyclone affecting the UK and Ireland. The blue arrows between isobarsindicate the direction of the wind, while the "L" symbol denotes the centre of the "low". Note the occluded, cold and warm frontal boundaries.

The descriptor "extratropical" refers to the fact that this type of cyclone generally occurs outside of the tropics, in the middle latitudes of the planet. These systems may also be described as "mid-latitude cyclones" due to their area of formation, or "post-tropical cyclones" where extratropical transition has occurred, and are often described as "depressions" or "lows" by weather forecasters and the public. These are the everyday phenomena which along with anti-cyclones, drive the weather over much of the Earth.

Although extratropical cyclones are almost always classified as baroclinic since they form along zones of temperature and dewpoint gradient within the westerlies, they can sometimes become barotropic late in their life cycle when the temperature distribution around the cyclone becomes fairly uniform with radius. An extratropical cyclone can transform into a subtropical storm, and from there into a tropical cyclone, if it dwells over warm waters and develops central convection, which warms its core.

Global Scale Meteorology

Global scale phenomena are weather patterns related to the transport of heat, wind, and moisture from the tropics to the poles. An important pattern is global atmospheric circulation, the large-scale movement of air that helps distribute thermal energy (heat) across the surface of the Earth.

Global atmospheric circulation is the fairly constant movement of winds across the globe. Winds develop as air masses move from areas of high pressure to areas of low pressure. Global atmospheric circulation is largely driven by Hadley cells. Hadley cells are tropical and equatorial convection patterns. Convection drives warm air high in the atmosphere, while cool, dense air pushes lower in a constant loop. Each loop is a Hadley cell.

Hadley cells determine the flow of trade winds, which meteorologists forecast. Businesses, especially those exporting products across oceans, pay close attention to the strength of trade winds because they help ships travel faster. Westerlies are winds that blow from the west in the midlatitudes. Closer to the Equator, trade winds blow from the northeast (north of the Equator) and the southeast (south of the Equator).

Meteorologists study long-term climate patterns that disrupt global atmospheric circulation. Meteorologists discovered the pattern of El Nino, for instance. El Niño involves ocean currents and trade winds across the Pacific Ocean. El Niño occurs roughly every five years, disrupting global atmospheric circulation and affecting local weather and economies from Australia to Peru.

El Niño is linked with changes in air pressure in the Pacific Ocean known as the Southern Oscillation. Air pressure drops over the eastern Pacific, near the coast of the Americas, while air pressure rises over the western Pacific, near the coasts of Australia and Indonesia. Trade winds weaken. Eastern Pacific nations experience extreme rainfall. Warm ocean currents reduce fish stocks, which depend on nutrient-rich upwelling of cold water to thrive. Western Pacific nations experience drought, devastating agricultural production.

Climatology

Climatology, branch of the atmospheric sciences concerned with both the description of climate and the analysis of the causes of climatic differences and changes and their practical consequences. Climatology treats the same atmospheric processes as meteorology, but it seeks as well to identify the slower-acting influences and longer-term changes of import, including the circulation of the oceans and the small yet measurable variations in the intensity of solar radiation.

From its origins in 6th-century-BC Greek science, climatology has developed along two main lines: regional climatology and physical climatology. The first is the study of discrete and characteristic weather phenomena of a particular continental or subcontinental region. The second involves a statistical analysis of the various weather elements, principally temperature, moisture, atmospheric pressure, and wind speed, and a detailed examination of the basic relationships between such elements. Since the 1960s a third main branch, dynamic meteorology, has emerged. It deals primarily with the numerical simulation of climate and climatic change, employing models of atmospheric processes based on the fundamental equations of dynamic meteorology. Other significant subdisciplines of climatology include bioclimatology and paleoclimatology.

Instruments of Climatology

Here are some examples of equipment used my climatologists on Earth and other planets. Some of you might be asking, "Aren't these the same things that the weatherman uses?" Yes. The big difference between the weatherman's needs and a climatologist's needs is timeliness. Weather forecasters only care about today, tomorrow, and next week. A climatologist uses readings from this equipment and stores the data for years and looks for long-term trends and patterns.

Barometer

Barometers are used to measure the pressure in the atmosphere. It's the device used to determine that the pressure is lower in Denver, Colorado than it is at sea level in Seattle, Washington. It also has helped meteorologists determine when weather changes are going to happen. You already learned that different pressures in the surrounding areas create different types of weather. A changing of pressure in the atmosphere also indicates that storm fronts may be approaching. Climatologists use barometric data over periods of months and years.

Hygrometer

This is a big word for the device that measures the humidity in the local atmosphere. Humidity is the percentage of water vapor in the surrounding air. You might know the difference in humidity on your own. Days with really low humidity sometimes make your lips chap and give you skin dry. High humidity days are often thick and you sweat easily.

Thermometer

Everyone has probably heard of these. Thermometers are used in many places to measure temperature. If you get sick, that's one type of thermometer. Another type is used to measure the temperature in your oven. Another is used in your car's radiator. Climatologists use their own that measures temperatures in a specific range.

Sub-Fields of Climatology

According to the area of specialization, climatology has been divided into smaller sub-fields: paleoclimatology, paleotempestology, historical climatology, metrology, and bioclimatology. Paleoclimatology focuses on establishing the past climatic patterns of a place by studying ice cores and tree rings. Paleotempestology uses ancient data to determine the frequency and magnitude of past hurricanes. Historical climatology focuses on establishing the climate of a place after studying the activities that the ancient dwellers of the particular place engaged in. Metrology, which is often confused with climatology, deals with weather, which runs for a maximum of probably a week or a month. Bioclimatology deals with the effects the climate has on the living organisms.

Importance of Climatology

Climatology is important in determining the climatic patterns of a particular region. Establishing the climatic pattern is significant in deciding the economic activities that would thrive in that particular region. If the climate of a region is established to be cool and wet, it would be safe to conclude that agriculture might thrive in the region. Having a clear climate pattern makes it easier for people to understand the seasons of engaging in particular tasks. This is especially most important to tourists and farmers. Infrastructure development, especially buildings are dependent on climate. After a climatic pattern has been established, the engineers recommend the use of materials that would not only withstand the conditions but also protect the dwellers from any harsh climatic condition. Furthermore, climatology seeks to establish why climate varies from place to another.

Dendroclimatology

The study of the annual growth of trees and the consequent assembling of long, continuous chronologies for use in dating wood is called dendrochronology. The study of the relationships between annual tree growth and climate is called dendroclimatology. Dendroclimatology offers a high resolution (annual) form of palaeoclimate reconstruction for most of the Holocene.

The annual growth of a tree is the net result of many complex and interrelated biochemical processes. Trees interact directly with the microenvironment of the leaf and the root surfaces. The fact that there exists a relationship between these extremely localised conditions and larger scale climatic parameters offers the potential for extracting some measure of the overall influence of climate on growth from year to year. Growth may be affected by many aspects of the microclimate: sunshine, precipitation, temperature, wind speed and humidity. Besides these, there are other non-climatic factors that may exert an influence, such as competition, defoliators and soil nutrient characteristics.

There are several subfields of dendroclimatology associated with the processing and interpretation of different tree-growth variables. Such variables include treering width, densitometric parameters and chemical or isotopic variables.

A cross section of most temperate forest tree trunks will reveal an alternation of lighter and darker bands, each of which is usually continuous around the tree circumference. These are seasonal growth increments produced by meristematic tissues in the cambium of the tree. Each seasonal increment consists of a couplet of earlywood (a light growth band from the early part of the growing season) and denser latewood (a dark band produced towards the end of the growing season), and collectively they make up the tree ring. The mean width of the tree ring is a function of many variables, including the tree species, tree age, soil nutrient availability, and a whole host of climatic factors. The problem facing the dendroclimatologist is to extract whatever climatic signal is available in the tree-ring data from the remaining background "noise".

Whenever tree growth is limited directly or indirectly by some climate variable, and that limitation can be quantified and dated, dendroclimatology can be used to reconstruct some information about past environmental conditions. Only for trees growing near the extremities of their ecological amplitude, where they may be subject to considerable climatic stresses, is it likely that climate will be a limiting factor. Commonly two types of climatic stress are recognised, moisture stress and temperature stress. Trees growing in semi-arid regions are frequently limited by the availability of water, and dendroclimatic indicators primarily reflect this variable. Trees growing near the latitudinal or altitudinal treeline are mainly under growth limitations imposed by temperature; hence dendroclimatic indicators in such trees contain strong temperature signals.

Furthermore, climatic conditions prior to the growth period may precondition physiological processes within the tree and hence strongly influence subsequent growth. Consequently, strong serial correlation or autocorrelation may establish itself in the tree-ring record. A specific tree ring will contain information not just about the climate conditions of the growth year but information about the months and years preceding it.

Several assumptions underlie the production of quantitative dendroclimatic reconstructions. First, the physical and biological processes which link toady's environment with today's variations in tree growth must have been in operation in the past. This is the principle of uniformitarianism. Second, the climate conditions which produce anomalies in tree-growth patterns in the past must have their analogue during the calibration period. Third, climate is continuous over areas adjacent to the domain of the tree-ring network, enabling the development of a statistical transfer function relating growth in the network to climate variability inside and outside of it. Finally, it is assumed that the systematic relationship between climate as a limiting factor and the biological response can be approximated as a linear mathematical expression. Fritts provides a more exhaustive review of the assumptions involved in the use of dendroclimatology.

The general approach taken in dendroclimatic reconstruction is:

1) To collect (sample) data from a set of trees (within a tree population) which have been selected on the basis that climate (e.g. temperature, moisture) should be a limiting factor.
2) To assemble the data into a composite site chronology by cross-dating the individual series after the removal of age effects by standardization. This master chronology increases the (climate) signal to (non-climate) noise ratio.
3) To build up a network of site chronologies for a region.
4) To identify statistical relationships between the chronology times series and the instrumental climate data for the recent period - the calibration period.
5) To use these relationships to reconstruct climatic information from the earlier period covered by the tree-ring data, and;
6) Finally, to test, or verify, the resulting reconstruction against independent data.

Bradley gives a full account of the methods (1 to 6 above) of palaeoclimate reconstruction from tree-ring analysis. This approach may be applied to all the climate-dependent tree-growth variables, specifically tree-ring width, but also wood density and isotopic measurements. The latewood of a tree ring is much denser than the earlywood and interannual variations contain a strong climatic signal. Density variations are particularly valuable in dendroclimatology because they to not change significantly with tree age, and the process of standardisation (removal of growth function) can therefore be avoided.

The use of isotopic measurements in dendroclimatology also avoids the need for a standardisation process. The basic premise of isotope dendroclimatology is that since $^{18}O/^{16}O$ and D/H (deuterium/hydrogen) variations in meteoric (atmospheric) waters are a function of temperature, tree growth which records such isotope variations should preserve a record of past temperature fluctuations. Unfortunately, isotope fractionation effects within the tree, which are themselves temperature dependent, will create problems associated with this technique.

Historical Climatology

Historical climatology is the study of historical changes in climate and their effect on human history and development. This differs from paleoclimatology which encompasses climate change over the entire history of Earth. The study seeks to define periods in human history where temperature or precipitation varied from what is observed in the present day. The primary sources include written records such as sagas, chronicles, maps and local history literature as well as pictorial representations such as paintings, drawings and even rock art. The archaeological record is equally important in establishing evidence of settlement, water and land usage.

Techniques of Historical Climatology

In literate societies, historians may find written evidence of climatic variations over hundreds or thousands of years, such as phenological records of natural processes, for example viticultural records of grape harvest dates. In preliterate or non-literate societies, researchers must rely on other techniques to find evidence of historical climate differences.

Past population levels and habitable ranges of humans or plants and animals may be used to find evidence of past differences in climate for the region. Palynology, the study of pollens, can show not only the range of plants and to reconstruct possible ecology, but to estimate the amount of precipitation in a given time period, based on the abundance of pollen in that layer of sediment or ice.

Evidence of Climatic Variations

The eruption of the Toba supervolcano, 70,000 to 75,000 years ago reduced the average global temperature by 5 degrees Celsius for several years and may have triggered an ice age. It has been postulated that this created a bottleneck in human evolution. A much smaller but similar effect occurred after the eruption of Krakatoa in 1883, when global temperatures fell for about 5 years in a row.

Before the retreat of glaciers at the start of the Holocene (~9600 BC), ice sheets covered much of the northern latitudes and sea levels were much lower than they are today. The start of our present interglacial period appears to have helped spur the development of human civilization.

Human Record

The 16th-century Skálholt Map of Norse America.

One of Grimspound's hut circles.

Evidence of a warm climate in Europe, for example, comes from archaeological studies of settlement and farming in the Early Bronze Age at altitudes now beyond cultivation, such as Dartmoor, Exmoor, the Lake district and the Pennines in Great Britain. The climate appears to have deteriorated towards the Late Bronze Age however. Settlements and field boundaries have been found at high altitude in these areas, which are now wild and uninhabitable. Grimspound on Dartmoor is well preserved and shows the standing remains of an extensive settlement in a now inhospitable environment.

Some parts of the present Saharan desert may have been populated when the climate was cooler and wetter, judging by cave art and other signs of settlement in Prehistoric Central North Africa.

The Medieval Warm Period was a time of warm weather between about AD 800–1300, during the European Medieval period. Archaeological evidence supports studies of the Norse sagas which describe the settlement of Greenland in the 9th century AD of land now quite unsuitable for cultivation. For example, excavations at one settlement site have shown the presence of birch trees during the early Viking period. The same period records the discovery of an area called Vinland, probably in North America, which may also have been warmer than at present, judging by the alleged presence of grape vines. The interlude is known as the Medieval Warm Period.

Little Ice Age

Later examples include the Little Ice Age, well documented by paintings, documents (such as diaries) and events such as the River Thames frost fairs held on frozen lakes and rivers in the 17th and 18th centuries. The River Thames was made more narrow and flowed faster after old London Bridge was demolished in 1831, and the river was embanked in stages during the 19th century, both of which made the river less liable to freezing:

> "And in this same year after the 2nd of February came the severe winter with frost and snow, and with all kinds of bad weather, so that there was no man alive who could remember so severe a winter as that, both through mortality of men and disease of cattle; both birds and fishes perished through the great cold and hunger."

The *Chronicle* is the single most important historical source for the period in England between the departure of the Romans and the decades following the Norman Conquest. Much of the information given in the *Chronicle* is not recorded elsewhere.

The Frozen Thames.

The Little Ice Age brought colder winters to parts of Europe and North America. In the mid-17th century, glaciers in the Swiss Alps advanced, gradually engulfing farms and crushing entire villages. The River Thames and the canals and rivers of the Netherlands often froze over during the winter, and people skated and even held frost fairs on the ice. The first Thames frost fair was in 1607; the last in 1814, although changes to the bridges and the addition of an embankment affected the river flow and depth, diminishing the possibility of freezes. The freeze of the Golden Horn and the southern section of the Bosphorus took place in 1622. In 1658, a Swedish army marched across the Great Belt to Denmark to invade Copenhagen. The Baltic Sea froze over, enabling sledge rides from Poland to Sweden, with seasonal inns built on the way. The winter of 1794/1795 was particularly harsh when the French invasion army under Pichegru could march on the frozen rivers of the Netherlands, while the Dutch fleet was fixed in the ice in Den Helder harbour. In the winter of 1780, New York Harbour froze, allowing people to walk from Manhattan to Staten Island. Sea ice surrounding Iceland extended for miles in every direction, closing that island's harbours to shipping.

The last written records of the Norse Greenlanders are from a 1408 marriage in Hvalsey Church — today the best-preserved of the Norse ruins.

The severe winters affected human life in ways large and small. The population of Iceland fell by half, but this was perhaps also due to fluorosis caused by the eruption of the volcano Laki in 1783. Iceland also suffered failures of cereal crops and people moved away from a grain-based diet. The Norse colonies in Greenland starved and vanished (by the 15th century) as crops failed and livestock could not be maintained through increasingly harsh winters, though Jared Diamond noted that they had exceeded the agricultural carrying capacity before then. In North America, American Indians formed leagues in response to food shortages. In Southern Europe, in Portugal, snow storms were much more frequent while today they are rare. There are reports of heavy snowfalls in the winters of 1665, 1744 and 1886.

In contrast to its uncertain beginning, there is a consensus that the Little Ice Age ended in the mid-19th century.

Evidence of Anthropogenic Climate Change

Through deforestation and agriculture, some scientists have proposed a human component in some historical climatic changes. Human-started fires have been implicated in the transformation of much of Australia from grassland to desert. If true, this would show that non-industrialized societies could have a role in influencing regional climate. Deforestation, desertification and the salinization of soils may have contributed to or caused other climatic changes throughout human history.

Paleoclimatology

Paleoclimatology is the study of ancient climates, prior to the widespread availability of instrumental records. Similar to the way archeologists study fossils and other physical clues to gain insight into the prehistoric past, paleoclimatologists study several different types of environmental evidence to understand what the Earth's past climate was like and why. Over the years, the Earth has kept records of its climate conditions preserved in tree rings, locked in the skeletons of tropical coral reefs, sealed in glaciers and ice caps, and buried in laminated sediments from lakes and the ocean. Scientists can use those environmental recorders to estimate past conditions, extending our understanding of climate back hundreds to millions of years.

Trees like the Giant Sequoia pictured here can grow to be hundreds or even thousands of years old, providing an important source of information about past environmental variations.

If there is one thing that the paleoclimate record shows, it's that the Earth's climate is always changing. Over the past two million years, numerous glacial periods have covered much of the high-latitude northern hemisphere landmasses in glacial ice, dropped sea level as much as 410 feet, and significantly cooled even tropical regions. In the more distant past, the Cretaceous Period (between 145.5 and 65.5 million years ago) was significantly warmer than today, with less polar ice, rising sea levels, and warm weather organisms thriving even in near-polar regions.

The study of paleoclimates has been particularly helpful in showing that the Earth's climate system can shift between dramatically different climate states in a matter of years or decades. For example, tree-ring and lake-sediment records from North America show that decadal-scale "megadroughts" occurred multiple times over the last thousand years. During these periods, persistent droughts lasted much longer than any of the droughts we have experienced over the period of instrumental records. Understanding "climate surprises" of the past is critical to avoid being surprised by abrupt climatic change in the future.

The study of past climate change also helps us understand how humans influence the Earth's climate system. The climatic record over the last thousand years clearly shows that global temperatures increased significantly in the 20th century, and that this warming was likely to have been unprecedented in the last 1,200 years. The paleoclimatic record also allows us to examine the causes of past climate change and to help unravel how much of the 20th century warming may be explained by natural causes.

Tornado Climatology

Tornado Climatology is a branch of Climatology which deals with the emergence, development and distribution of tornadoes. It solves with such issues as the possible origins of tornadoes (which has not yet been thoroughly studied), predictions of their movement.

Tornado is a vortex that arises in a thundercloud and spreads downward, often to the very surface of the earth, in the form of a sleeve or trunk made of clouds.

The causes of tornadoes can be different - some tornadoes arise because of difference between warm air and cold sea water. A tornado exists because of the different pressure caused by the difference in temperatures. Inside the tornado warm air rises upward, and that is why a zone of rarefied air is formed, into which the nearest air and clouds rush. Due to this process, the tornado gains size and power sharply and avalanche-like. Then the tornado moves to the side where cold air comes from. The tornado can lift large and tattered objects, and also destroy buildings by explosion because of the difference in pressure in the building and in the tornado.

Urban Climatology

Urban climatology refers to a specific branch of climatology that is concerned with interactions between urban areas and the atmosphere, the effects they have on one another, and the varying spatial and temporal scales at which these processes (and responses) occur.

Influential Factors

Several factors influence the urban climate, including city size, the morphology of the city, land-use configuration, and the geographic setting (such as relief, elevation, and regional climate). Some of the differences between urban and rural climates include air quality, wind patterns, and changes in rainfall patterns, but one of the most studied is the urban heat island effect.

Temperature and Urban Heat Island Effect

Urban environments are typically warmer than their surroundings, as documented over a century ago by Howard. Urban areas are islands or spots on the broader scale compared with more rural surrounding land. The spatial distribution of temperatures occurs in tandem with temporal changes, which are both causally related to anthropogenic sources.

The urban environment has two atmosphere layers, besides the planetary boundary layer outside and extending well above the city: (1) The urban boundary layer is due to the spatially integrated heat and moisture exchanges between the city and its overlying air. (2) The surface of the city corresponds to the level of the urban canopy layer. Fluxes across this plane comprise those from individual units, such as roofs, canyon tops, trees, lawns, and roads, integrated over larger land-use divisions (for example, suburbs). The urban heat island effect has been a major focus of urban climatological studies, and in general the effect the urban environment has on local meteorological conditions.

Pollution

The field also includes the topics of air quality, Radiation Fluxes, Micro-Climates and even issues traditionally associated with architectural design and engineering, such as Wind Engineering. Causes and effects of pollution as understood through Urban Climatology are becoming more important for Urban Planning.

Precipitation

Changes in winds and convection patterns over and around cities impacts precipitation. Contributing factors are believed to be urban heat island, heightened surface roughness, and increased aerosol concentration.

Climate Change

Urban climatology is strongly linked to research surrounding global warming. As centers for socio-economic activities, cities produce large amounts of greenhouse gases, most notably CO_2 as a consequence of human activities such as transport, development, waste related to heating and cooling requirements etc.

Globally, cities are expected to grow into the 21st century (and beyond) - as they grow and develop the landscapes in which they inhabit will change so too will the atmosphere resting above them, increasing emissions of GHG's thus contributing to the global green house effect.

Finally, many cities are vulnerable to the projected consequences of climate change (sea level rise, changes in temperature, precipitation, storm frequency) as most develop on or near coastlines, nearly all produce distinct urban heat islands and atmospheric pollution: as areas in which there is concentrated human habitation these effects potentially will have the largest and most dramatic impact (e.g. France's heat wave in 2003) and thus are a major focus for urban climatology.

Spatial Planning and Public Health

Urban Climatology impacts decision-making for municipal planning and policy in regards to pollution, extreme heat events, and stormwater modeling.

Tropical Cyclone Climatology

A tropical cyclone is a rotating, organized system of clouds and thunderstorms that originates over tropical or subtropical waters and has a closed low-level circulation. Tropical cyclones rotate counterclockwise in the Northern Hemisphere. They are classified as follows:

- Tropical Depression: A tropical cyclone with maximum sustained winds of 38 mph (33 knots) or less.
- Tropical Storm: A tropical cyclone with maximum sustained winds of 39 to 73 mph (34 to 63 knots).
- Hurricane: A tropical cyclone with maximum sustained winds of 74 mph (64 knots) or higher. In the western North Pacific, hurricanes are called typhoons; similar storms in the Indian Ocean and South Pacific Ocean are called cyclones.
- Major Hurricane: A tropical cyclone with maximum sustained winds of 111 mph (96 knots)

or higher, corresponding to a Category 3, 4 or 5 on the Saffir-Simpson Hurricane Wind Scale.

Tropical cyclones forming between 5 and 30 degrees North latitude typically move toward the west. Sometimes the winds in the middle and upper levels of the atmosphere change and steer the cyclone toward the north and northwest. When tropical cyclones reach latitudes near 30 degrees North, they often move northeast.

Tropical Cyclone Rainfall Climatology

A tropical cyclone rainfall climatology is developed to determine rainfall characteristics of past tropical cyclones. A tropical cyclone rainfall climatology can be used to help forecast current or upcoming tropical cyclone impacts. The degree of a tropical cyclone rainfall impact depends upon speed of movement, storm size, and degree of vertical wind shear. One of the most significant threats from tropical cyclones is heavy rainfall. Large, slow moving, and non-sheared tropical cyclones produce the heaviest rains. The intensity of a tropical cyclone appears to have little bearing on its potential for rainfall over land, but satellite measurements over the last several years show that more intense tropical cyclones produce noticeably more rainfall over water. Flooding from tropical cyclones remains a significant cause of fatalities, particularly in low-lying areas.

Anticipating a Flood Event

While inland flooding is common to tropical cyclones, there are factors which lead to excessive rainfall from tropical cyclones. Slow motion, as was seen during Hurricane Danny (1997) and Hurricane Wilma, can lead to high amounts of rainfall. The presence of mountains/hills near the coast, like across much of Mexico, Haiti, the Dominican Republic, Central America, Madagascar, Réunion, China, and Japan acts to magnify rainfall potential due to forced upslope flow into the mountains. Strong upper level forcing from a trough moving through the Westerlies and its associated cold front, as was the case during Hurricane Floyd, can lead to high amounts even from systems moving at an average forward motion. Larger tropical cyclones drop more rainfall as they precipitate upon one spot for a longer time frame than average or small tropical cyclones. A combination of two of these factors could be especially crippling, as was seen during Hurricane Mitch in Central America. During the 2005 season, flooding related to slow-moving Hurricane Stan's broad circulation led to 1,662–2,000 deaths.

General Distribution within a Tropical Cyclone

Rainfall Rate per day within radius of the center (Riehl)			
Radius (mi)	Radius (km)	Amount (in)	Amount (mm)
35	56	33.98	863
70	112	13.27	337
140	224	4.25	108
280	448	1.18	30

Isaac Cline was the first to investigate rainfall distribution around tropical cyclones in the early

1900s. He found that a larger proportion of rainfall falls in advance of the center (or eye) than after the center's passage, with the highest percentage falling in the right front quadrant. Father Viñes of Cuba found that some tropical cyclones have their highest rainfall rates in the rear quadrant within a training (non-moving) inflow band. Normally, as a tropical cyclone intensifies, its heavier rainfall rates become more concentrated around its center. Rainfall is found to be heaviest in tropical cyclone's inner core, whether it be the eyewall or central dense overcast, within a degree latitude of the center, with lesser amounts farther away from the center. Most of the rainfall in tropical cyclones is concentrated within its radius of gale-force (34 knots/39 mph/63 km/h) winds. Rainfall is more common near the center of tropical cyclones overnight. Over land, outer bands are more active during the heating of the day, which can act to restrict inflow into the center of the cyclone. Recent studies have shown that half of the rainfall within a tropical cyclone is stratiform in nature. The chart to the right was developed by Riehl in 1954 using meteorological equations that assume a gale radius of about 140 miles (230 km), a fairly symmetric cyclone, and does not consider topographic effects or vertical wind shear. Local amounts can exceed this chart by a factor of two due to topography. Wind shear tends to lessen the amounts below what is shown on the table.

Relation to Storm Size

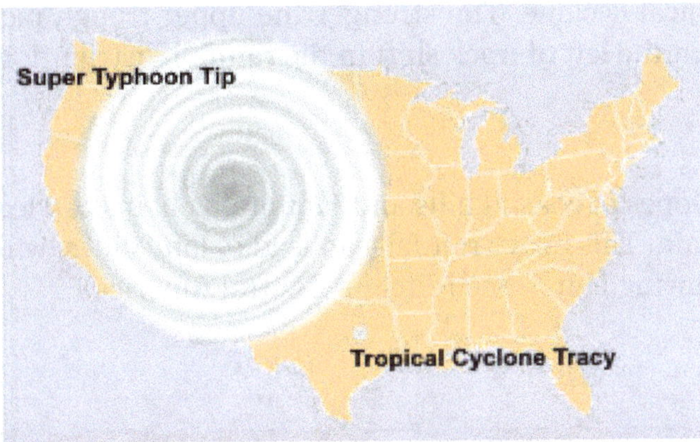

The relative sizes of Typhoon Tip, Cyclone Tracy, and the United States.

Larger tropical cyclones have larger rain shields, which can lead to higher rainfall amounts farther from the cyclone's center. This is generally due to the longer time frame rainfall falls at any one spot in a larger system, when compared to a smaller system. Some of the difference seen concerning rainfall between larger and small storms could be the increased sampling of rainfall within a larger tropical cyclone when compared to that of a compact cyclone; in other words, the difference could be the result of a statistical problem.

Slow/looping Motion on Rainfall Magnitude

Storms which have moved slowly, or loop, over a succession of days lead to the highest rainfall amounts for several countries. Riehl calculated that 33.97 inches (863 mm) of rainfall per day can be expected within one-half degree, or 35 miles (56 km), of the center of a mature tropical cyclone. Many tropical cyclones progress at a forward motion of 10 knots, which would limit the duration of this excessive rainfall to around one-quarter of a day, which would yield about 8.50 inches (216 mm) of rainfall. This would be true over water, within 100 miles (160 km) of the coastline, and outside topographic features. As a cyclone moves farther inland and is cut off from its supply of warmth and moisture (the ocean), rainfall amounts from tropical cyclones and their remains decrease quickly.

Vertical Wind Shear Impact on Rainfall Shield

Vertical wind shear forces the rainfall pattern around a tropical cyclone to become highly asymmetric, with most of the precipitation falling to the left and downwind of the shear vector, or downshear left. In other words, southwesterly shear forces the bulk of the rainfall north-northeast of the center. If the wind shear is strong enough, the bulk of the rainfall will move away from the center leading to what is known as an exposed circulation center. When this occurs, the potential magnitude of rainfall with the tropical cyclone will be significantly reduced.

Effect of Interaction with Frontal Boundaries/Upper Level Troughs

As a tropical cyclone interacts with an upper-level trough and the related surface front, a distinct northern area of precipitation is seen along the front ahead of the axis of the upper level trough. This type of interaction can lead to the appearance of the heaviest rainfall falling along and to the left of the tropical cyclone track, with the precipitation streaking hundreds of miles or kilometers downwind from the tropical cyclone. The stronger the upper trough picking up the tropical cyclone, the more significant the left of track shift in the rainfall distribution tends to be.

Mountains

Moist air forced up the slopes of coastal hills and mountain chains can lead to much heavier rainfall than in the coastal plain. This heavy rainfall can lead to landslides, which still cause significant loss of life such as seen during Hurricane Mitch in Central America.

Global Distribution

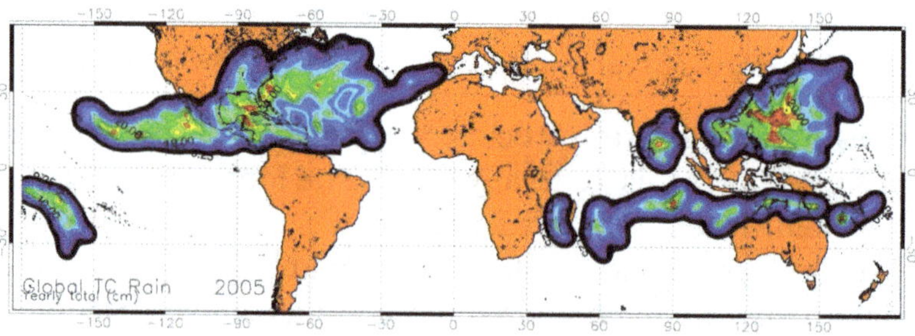

Global tropical cyclone rainfall in 2005.

Globally, tropical cyclone rainfall is more common across the northern hemisphere than across the southern hemisphere. This is mainly due to the normal annual tropical cyclone distribution, as between half and two-thirds of all tropical cyclones form north of the equator. Rainfall is concentrated near the 15th parallel in both hemispheres, with a less steep dropoff seen with latitude across the northern hemisphere, due to the stronger warm water currents seen in that hemisphere which allow tropical cyclones to remain tropical in nature at higher latitudes than south of the equator. In the southern hemisphere, rainfall impacts will be most common between January and March, while north of the equator, tropical cyclone rainfall impacts are more common between June and November. Japan receives over half of its rainfall from typhoons.

United States Tropical Cyclone Rainfall Statistics

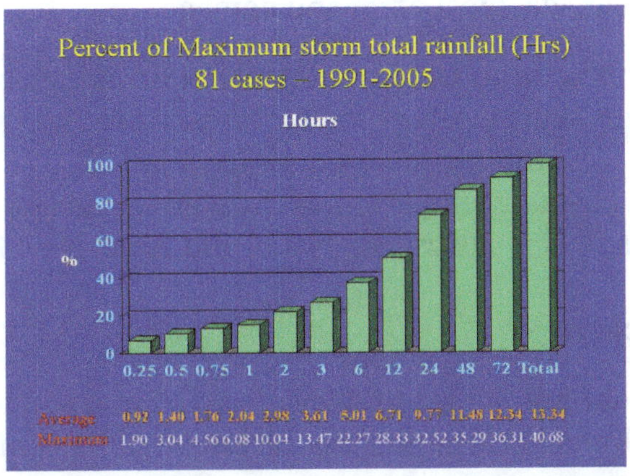

U.S. Tropical Cyclone Rainfall Accumulations per time frame.

Between 1970-2004, inland flooding caused a majority of the tropical cyclone-related fatalities in the United States. This statistic changed in 2005, when Hurricane Katrina's impact alone shifted the most deadly aspect of tropical cyclones back to storm surge, which has historically been the most deadly aspect of strong tropical cyclones. On average, five tropical cyclones of at least tropical depression strength lead to rainfall across the contiguous United States annually, contributing around a quarter of the annual rainfall to the southeast United States. While many of these storms form in the Atlantic basin, some systems or their remnants move through Mexico from the Eastern Pacific basin. The average storm total rainfall for a tropical cyclone impacting the lower 48 from the Atlantic basin is about 16 inches (406 mm), with 70–75 percent of the storm total falling within a 24-hour period. The highest point total was seen during Hurricane Harvey in 2017, when 60.58 inches (1,538.7 mm) fell in southeast Texas.

Tropical Cyclone Windspeed Climatology

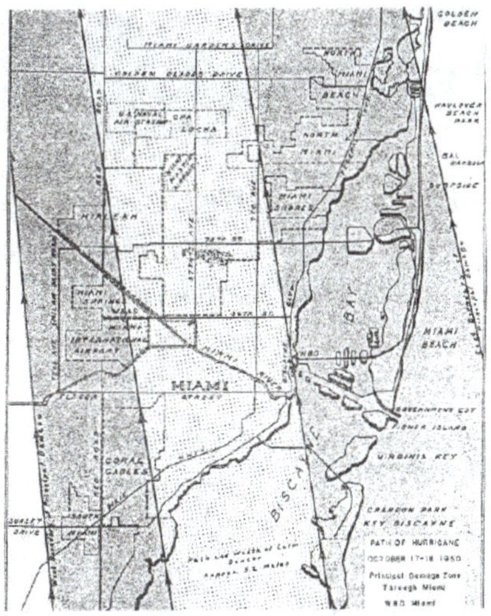

Wind Swath of Hurricane King over Miami.

Tropical cyclone windspeed climatology is the study wind distribution amongst tropical cyclones, a significant threat to land and people. Since records began in 1851, winds from hurricanes, typhoons and cyclones have been responsible for fatalities and damage in every basin. Major hurricanes

(Category 3 or above) usually cause the most wind damage. Hurricane Andrew for example caused $45 billion (2005 USD) in damage, most of it wind damage.

Although wind damage is common to tropical cyclones near landmasses, there are a few factors that lead to high wind speeds. Warm water temperatures, which was seen during Hurricane Wilma when it winds rapidly strengthened to 185 mph (295 km/h) in a 24-hour period due to the presence of abnormally warm water temperatures. Size and speed of the storm, which results in damage along a wide area as seen in Hurricane Isabel and Great New England Hurricane.

Winds in Tropical Cyclones

The winds in a tropical cyclone are the result of evaporation and condensation of moisture which results in updrafts. The updrafts in turn increase the height of the storm which causes more condensation.

Location of the Winds

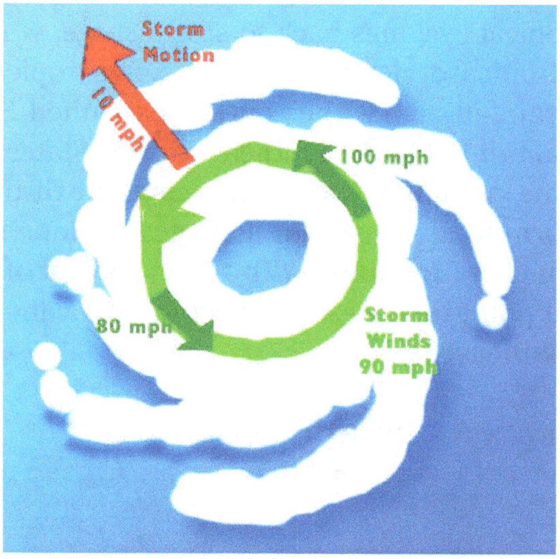

Right front quadrant of a typical hurricane.

The strongest winds in a northern hemisphere tropical cyclone is located in the eyewall and the right front quadrant of the tropical cyclone. Severe damage is usually the result when the eyewall of a hurricane, typhoon or cyclone passes over land. The right front quadrant is also an area of a tropical cyclone were the winds are strongest. The reason that the winds are at the front right side of a storm in the northern hemisphere (and the front left hand side in the Southern Hemisphere) is because of the motion of a tropical cyclones motion contributing to its rotation. A 100 mph hurricane while stationary, might have winds of 100 mph on its right side in the northern hemisphere while the rest of the storm might have winds at 70 mph. The location of the right (or left in the Southern Hemisphere) front quadrant also depends on the storm track. For example, in the northern hemisphere, if the storm was moving west, then the right side is to the north, if it moving north, then the strongest winds will be to the east of the center.

In the southern hemisphere, the strongest winds are to the left of the eye. That is because cyclonic winds below the equator, spin clockwise.

On occasion, strong winds and wind gusts can occur in the rain bands of a tropical cyclone. And inside the eye the winds are relatively calm. At higher altitudes, the winds within a tropical cyclone head away from the center, forming the outflow which produces the shape of the tropical cyclone.

Classification of the Winds

Saffir–Simpson scale						
TD	TS	C1	C2	C3	C4	C5

The United States uses the Saffir–Simpson hurricane wind scale, a 1–5 scale which categorize hurricanes by sustained wind speed. The hurricane scale was created in 1969 by Herbert Saffir and Bob Simpson. The National Hurricane Center classifies hurricanes category 3 or above, major hurricanes. The Joint Typhoon Warning Center also uses a scale similar to the Saffir-Simpson Hurricane Scale. However, the JTWC classified typhoons with winds of at 150 mph or higher *Supertyphoons*.

The Bureau of Meteorology in Australia uses a 1–5 scale called tropical cyclone severity categories. Unlike the Saffir–Simpson Hurricane Scale, it categorizes tropical cyclones based on wind gusts.

Beaufort Wind Scale

Before the 1–5 scale was created in 1969 by the National Hurricane Center and later by the Bureau of Meteorology in Australia. Many tropical cyclones were simply ranked by the Beaufort Wind Scale which was created in the early 1800s by Francis Beaufort. The purpose of the scale was to standardize wind reports in ship logs. The scale was divided down into 14 forces (force 0 is calm while force 12 is hurricane).

Fastest Windspeed Per Basin

This lists the fastest wind speed of a tropical cyclone per basin since 1950 (NOTE: All wind speeds are 1-minute sustained and are in mph).

- North Atlantic: 190 mph (1980).
- East Pacific: 215 mph (2015).
- Central Pacific: 175 mph (1994).
- West Pacific: 215 mph (1961).
- North Indian: 165 mph (2007).
- South Indian: 165 mph (1994).
- Australia: 180 mph (2006).
- South Pacific: 185 mph (2002).
- South Atlantic: 105 mph (2004).

Windspeed Per Country

Australia

In December 1998, Cyclone Thelma produced wind gusts of 108 mph (174 km/h) and sustained winds of 80 mph (139 km/h) as it made landfall in Western Australia. Three months later, Cyclone Vance produced a record gust of 166 mph (267 km/h). Sustained winds of 113 mph (182 km/h) in Onslow.

Cyclone John made landfall near Port Hedland, Australia as a category 4 tropical cyclone. Port Hedland experienced winds of 78 mph (124 km/h) for a period of 18 hours. At Karratha, the winds were gusting to 70 mph (113 km/h). Cape Lambert experienced sustained winds of 95 mph (150 km/h) for a five-hour period with gusts reaching 130 mph (210 km/h).

In March 2000, Tropical Storm Tessi made landfall near Townsville, Queensland,Australia. A weather station recorded the towns record wind gust of 81 mph (130 km/h).

Canada

A fast moving Hurricane Hazel brought sustained winds of 77 mph (124 km/h) and wind gusts up to 90 mph (150 km/h). The remnants of Hurricane Audrey brought 80 mph winds across southern Canada. A weather station on Nova Scotia recorded 100 mph sustained winds when Hurricane Juan made landfall.

Japan

During Typhoon Vera's impact on Japan, weather station in Iwo Jima reported winds of 89 mph (143 km/h).

United States

Florida

During Hurricane Wilma's passage through Florida a weather station recorded sustained wind-speeds of 77 knots, gusting to 101 kts.

Dynamic Climatology

Dynamic climatology is the study of climate as a branch of physics. The climate system is a physical system governed by reasonably well-known physical laws (fluid dynamics, thermodynamics, radiative transfer, etc). There are 5 major components of the climate system: atmosphere, hydrosphere (ie, oceans, lakes), land surface, cryosphere (sea ice, permanent ice caps, snow) and biosphere. The global system is interrelated in many complex ways, and the study of dynamic climatology requires expertise from many branches of science and data from all parts of the globe.

Some major aspects of the climate system may be appreciated by considering simpler, idealized systems. A nonrotating Earth with no topography or oceans would have a climate far different from that observed. If the atmosphere were not permitted to move in this idealized case, the resulting climate would not vary with longitude, but only with latitude; eg, temperatures would be high at equatorial latitudes, low at poleward latitudes. The resulting temperature structure at the surface, at various levels in the atmosphere and at various latitudes could be calculated using the equations of thermodynamics and radiative transfer. If the atmosphere were allowed to move on this nonrotating Earth, the warm air in equatorial regions would rise, flow polewards, cool, sink and flow equatorwards once more, giving rise to a Hadley circulation. The temperature difference from pole to equator provides the driving force for the motions and would be reduced as a consequence of the flow.

If the planet now began to rotate, the nature of the flow (hence, the climate) would differ remarkably. A simple Hadley circulation would no longer be found but rather a flow from west to east at middle latitudes and a flow from east to west at lower latitudes. Superimposed on this general flow would be large ripples and eddies. Without the oceans, winter cold and summer warmth would be extreme and the coldest and warmest times of the year would occur at the solstices. If moisture were introduced into the system as oceans and water vapour in the air, the resulting climate would again change dramatically; changes would include the appearance of major east-west asymmetries in the climate. The oceans have a profound effect on climate through storage and transport of heat. Topography, ice and snow, etc, add further degrees of complexity to the system.

Physical Climatology

Physical climatology is the science of examining and modelling numerically the processes, magnitudes and directions of energy and mass exchanges within the atmosphere, as well as at and immediately below the Earth's surface. These factors are important in determining all scales of climate, from local to global and from diurnal to millennial.

Variability in the absorption of solar radiation by the Earth-atmosphere system is the driving force behind climate. Annually the system absorbs about 70% of the incoming solar radiation and emits a similar amount of infrared radiation to space. Small but cumulative changes to this quasi-equilibrium due to, for example, changes in solar emission, the Earth's orbital geometry or atmospheric composition can initiate chain reactions (feedback processes) that amount to climatic change. The effects of variations in orbital geometry are negligible in the short term, but on geological time scales, orbital variations produce changes to solar irradiance patterns that are believed to be the cause of ICE AGES. On the other hand, altering the atmosphere's ability to emit and transmit thermal radiation through the burning of fossil fuels (enhanced greenhouse effect) could force unprecedented rates of climatic change in the coming century.

Currently, the portion of the Earth equatorward of 40° latitude is an energy source, since radiant energy gains exceed losses here; poleward of 40° latitude, however, the system is an energy sink, where losses exceed gains. WINDS and OCEAN CURRENTS transport energy poleward to rectify this latitudinal imbalance, thereby ameliorating surface temperature differences between the tropics and the poles. Moreover, the Earth's surface has an annual surplus of radiant energy whereas the atmosphere has a deficit. This imbalance is caused largely by the thermal insulating effect of atmospheric water vapour, the Earth's most abundant and important greenhouse gas. Convective transport of heat from the surface prevents excessive surface warming and atmospheric cooling. Likewise, on a global basis, CLOUDS serve to cool the system, as their solar reflecting properties outweigh their thermal insulating properties.

References

- Hebbert, M.; Mackillop, F. (2013). "Urban Climatology Applied to Urban Planning: A Postwar Knowledge Circulation Failure". International Journal of Urban and Regional Research. 37: 1542–1558. doi:10.1111/1468-2427.1206
- Meteorology: sciencedaily.com, Retrieved 14 May 2020
- Dr. DeCaria (2005-12-07). "ESCI 241 – Meteorology; Lesson 16 – Extratropical Cyclones". Department of Earth Sciences, Millersville University, Millersville, Pennsylvania. Archived from the original on 2006-09-03.
- What-is-climatology: worldatlas.com, Retrieved 15 June 2020
- *Miller GH, Fogel ML, Magee JW, Gagan MK, Clarke SJ, Johnson BJ (July 2005).* "Ecosystem Collapse in Pleistocene Australia and a Human Role in Megafaunal Extinction". *Science. 309(5732): 287–290.* Bibcode*:*2005Sci...309..287M. doi*:*10.1126/science.1111288. PMID 16002615
- What-paleoclimatology: ncdc.noaa.gov, Retrieved 28 July 2020

Chapter 7

Sub-Fields of Earth Science

Many sciences like geophysics, geochemistry, geobiology and geoinfomatics are branches of earth science. They study the various chemical, physical and biological phenomena associated with the Earth. The topics elaborated in this chapter will help in gaining a better perspective of all the related aspects of earth and planetary science and their applications.

Geophysics

Geophysics is a subject of natural science concerned with the physical processes and physical properties of the Earth and its surrounding space environment, and the use of quantitative methods for their analysis. The term *geophysics* sometimes refers only to the geological applications: Earth's shape; its gravitational and magnetic fields; its internal structure and composition; its dynamics and their surface expression in plate tectonics, the generation of magmas, volcanism and rock formation. However, modern geophysics organizations use a broader definition that includes the water cycle including snow and ice; fluid dynamics of the oceans and the atmosphere; electricity and magnetism in the ionosphere and magnetosphere and solar-terrestrial relations; and analogous problems associated with the Moon and other planets.

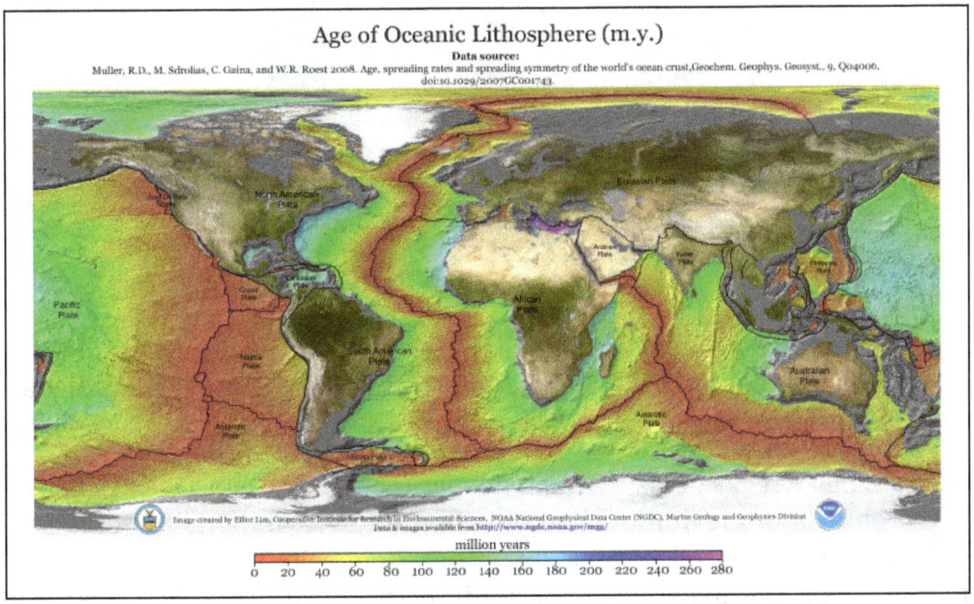

Age of the sea floor. Much of the dating information comes from magnetic anomalies.

Although geophysics was only recognized as a separate discipline in the 19th century, its origins date back to ancient times. The first magnetic compasses were made from lodestones, while more modern magnetic compasses played an important role in the history of navigation. The first seismic instrument was built in 132 BC. Isaac Newton applied his theory of mechanics to the tides and the precession of the equinox; and instruments were developed to measure the Earth's shape, density and gravity field, as well as the components of the water cycle. In the 20th century, geophysical methods were developed for remote exploration of the solid Earth and the ocean, and geophysics played an essential role in the development of the theory of plate tectonics.

Geophysics is applied to societal needs, such as mineral resources, mitigation of natural hazards and environmental protection. Geophysical survey data are used to analyze potential petroleum reservoirs and mineral deposits, locate groundwater, find archaeological relics, determine the thickness of glaciers and soils, and assess sites for environmental remediation.

Physical Phenomena

Geophysics is a highly interdisciplinary subject, and geophysicists contribute to every area of the Earth sciences. To provide a clearer idea of what constitutes geophysics, this section describes phenomena that are studied in physics and how they relate to the Earth and its surroundings.

Gravity

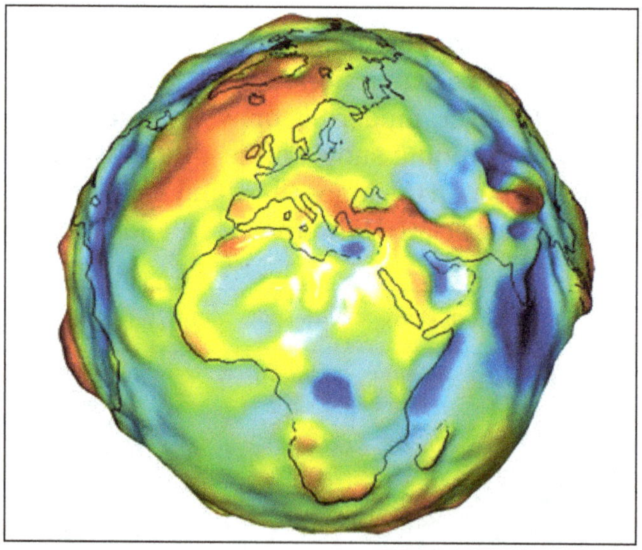

A map of deviations in gravity from a perfectly smooth, idealized Earth.

The gravitational pull of the Moon and Sun give rise to two high tides and two low tides every lunar day, or every 24 hours and 50 minutes. Therefore, there is a gap of 12 hours and 25 minutes between every high tide and between every low tide.

Gravitational forces make rocks press down on deeper rocks, increasing their density as the depth increases. Measurements of gravitational acceleration and gravitational potential at the Earth's surface and above it can be used to look for mineral deposits. The surface gravitational field provides information on the dynamics of tectonic plates. The geopotential surface called the geoid is one definition of the shape of the Earth. The geoid would be the global mean sea level if the oceans were in equilibrium and could be extended through the continents (such as with very narrow canals).

Heat Flow

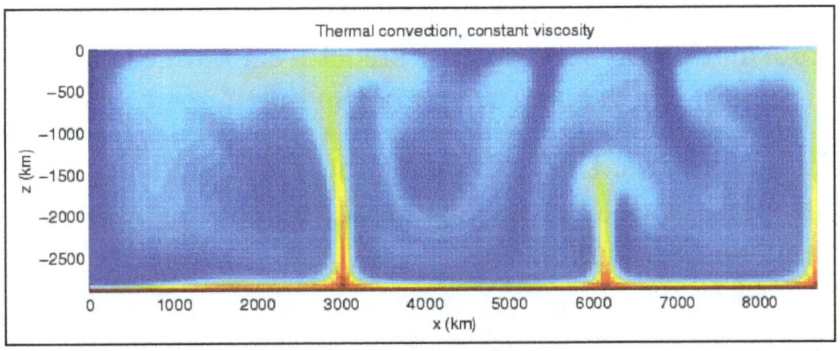

A model of thermal convection in the Earth's mantle. The thin red columns are mantle plumes.

The Earth is cooling, and the resulting heat flow generates the Earth's magnetic field through the geodynamo and plate tectonics through mantle convection. The main sources of heat are the primordial heat and radioactivity, although there are also contributions from phase transitions. Heat is mostly carried to the surface by thermal convection, although there are two thermal boundary layers – the core-mantle boundary and the lithosphere – in which heat is transported by conduction. Some heat is carried up from the bottom of the mantle by mantle plumes. The heat flow at the Earth's surface is about 4.2 × 1013 W, and it is a potential source of geothermal energy.

Vibrations

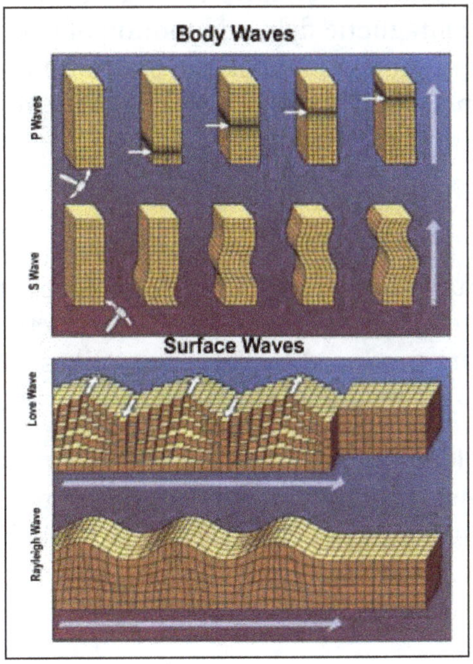

Illustration of the deformations of a block by body waves and surface waves.

Seismic waves are vibrations that travel through the Earth's interior or along its surface. The entire Earth can also oscillate in forms that are called normal modes or free oscillations of the Earth. Ground motions from waves or normal modes are measured using seismographs. If the waves come from a localized source such as an earthquake or explosion, measurements at more than one location can be used to locate the source. The locations of earthquakes provide information on plate tectonics and mantle convection.

Measurements of seismic waves are a source of information on the region that the waves travel through. If the density or composition of the rock changes suddenly, some waves are reflected. Reflections can provide information on near-surface structure. Changes in the travel direction, called refraction, can be used to infer the deep structure of the Earth.

Earthquakes pose a risk to humans. Understanding their mechanisms, which depend on the type of earthquake (e.g., intraplate or deep focus), can lead to better estimates of earthquake risk and improvements in earthquake engineering.

Electricity

Although we mainly notice electricity during thunderstorms, there is always a downward electric field near the surface that averages 120 V m−1. Relative to the solid Earth, the atmosphere has a net positive charge due to bombardment by cosmic rays. A current of about 1800 A flows in the global circuit. It flows downward from the ionosphere over most of the Earth and back upwards through thunderstorms. The flow is manifested by lightning below the clouds and sprites above.

A variety of electric methods are used in geophysical survey. Some measure spontaneous potential, a potential that arises in the ground because of man-made or natural disturbances. Telluric currents flow in Earth and the oceans. They have two causes: electromagnetic induction by the time-varying, external-origin geomagnetic field and motion of conducting bodies (such as seawater) across the Earth's permanent magnetic field. The distribution of telluric current density can be used to detect variations in electrical resistivity of underground structures. Geophysicists can also provide the electric current themselves.

Electromagnetic Waves

Electromagnetic waves occur in the ionosphere and magnetosphere as well as the Earth's outer core. Dawn chorus is believed to be caused by high-energy electrons that get caught in the Van Allen radiation belt. Whistlers are produced by lightning strikes. Hiss may be generated by both. Electromagnetic waves may also be generated by earthquakes.

In the Earth's outer core, electric currents in the highly conductive liquid iron create magnetic fields by electromagnetic induction. Alfvén waves are magnetohydrodynamic waves in the magnetosphere or the Earth's core. In the core, they probably have little observable effect on the geomagnetic field, but slower waves such as magnetic Rossby waves may be one source of geomagnetic secular variation.

Electromagnetic methods that are used for geophysical survey include transient electromagnetics and magnetotellurics.

Magnetism

The Earth's magnetic field protects the Earth from the deadly solar wind and has long been used for navigation. It originates in the fluid motions of the Earth's outer core. The magnetic field in the upper atmosphere gives rise to the auroras.

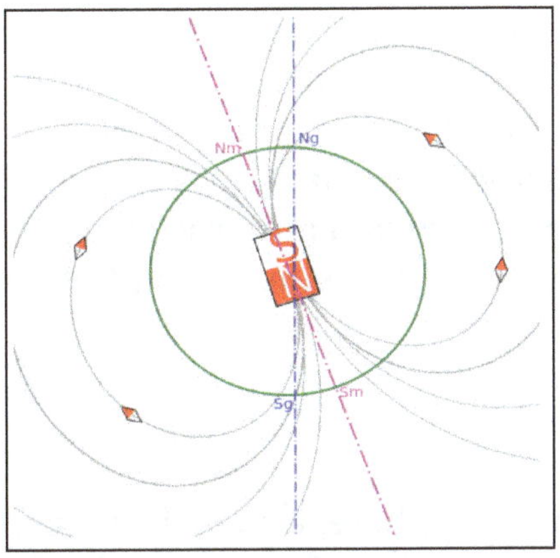

Earth's dipole axis (pink line) is tilted away from the rotational axis (blue line).

The Earth's field is roughly like a tilted dipole, but it changes over time (a phenomenon called geomagnetic secular variation). Mostly the geomagnetic pole stays near the geographic pole, but at random intervals averaging 440,000 to a million years or so, the polarity of the Earth's field reverses. These geomagnetic reversals, analyzed within a Geomagnetic Polarity Time Scale, contain 184 polarity intervals in the last 83 million years, with change in frequency over time, with the most recent brief complete reversal of the Laschamp event occurring 41,000 years ago during the last glacial period. Geologists observed geomagnetic reversal recorded in volcanic rocks, through magnetostratigraphy correlation and their signature can be seen as parallel linear magnetic anomaly stripes on the seafloor. These stripes provide quantitative information on seafloor spreading, a part of plate tectonics. They are the basis of magnetostratig-raphy, which correlates magnetic reversals with other stratigraphies to construct geologic time scales. In addition, the magnetization in rocks can be used to measure the motion of continents.

Radioactivity

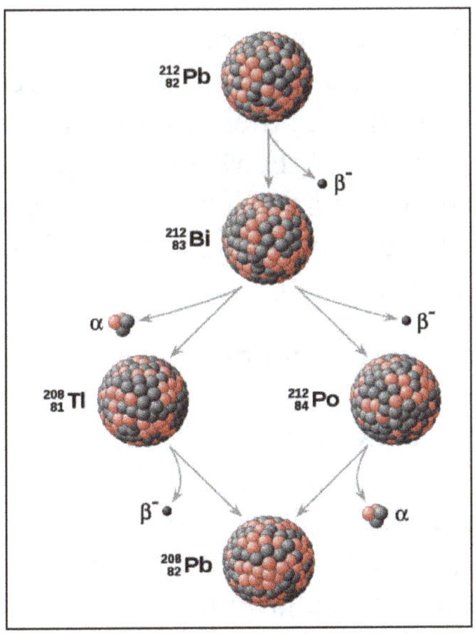

Example of a radioactive decay chain.

Radioactive decay accounts for about 80% of the Earth's internal heat, powering the geodynamo and plate tectonics. The main heat-producing isotopes are potassium-40, uranium-238, uranium-235, and thorium-232. Radioactive elements are used for radiometric dating, the primary method for establishing an absolute time scale in geochronology. Unstable isotopes decay at predictable rates, and the decay rates of different isotopes cover several orders of magnitude, so radioactive decay can be used to accurately date both recent events and events in past geologic eras. Radiometric mapping using ground and airborne gamma spectrometry can be used to map the concentration and distribution of radioisotopes near the Earth's surface, which is useful for mapping lithology and alteration.

Fluid Dynamics

Fluid motions occur in the magnetosphere, atmosphere, ocean, mantle and core. Even the mantle, though it has an enormous viscosity, flows like a fluid over long time intervals. This flow is reflected in phenomena such as isostasy, post-glacial rebound and mantle plumes. The mantle flow drives plate tectonics and the flow in the Earth's core drives the geodynamo.

Geophysical fluid dynamics is a primary tool in physical oceanography and meteorology. The rotation of the Earth has profound effects on the Earth's fluid dynamics, often due to the Coriolis effect. In the atmosphere it gives rise to large-scale patterns like Rossby waves and determines the basic circulation patterns of storms. In the ocean they drive large-scale circulation patterns as well as Kelvin waves and Ekman spirals at the ocean surface. In the Earth's core, the circulation of the molten iron is structured by Taylor columns.

Waves and other phenomena in the magnetosphere can be modeled using magnetohydrodynamics.

Mineral Physics

The physical properties of minerals must be understood to infer the composition of the Earth's interior from seismology, the geothermal gradient and other sources of information. Mineral physicists study the elastic properties of minerals; their high-pressure phase diagrams, melting points and equations of state at high pressure; and the rheological properties of rocks, or their ability to flow. Deformation of rocks by creep make flow possible, although over short times the rocks are brittle. The viscosity of rocks is affected by temperature and pressure, and in turn determines the rates at which tectonic plates move.

Water is a very complex substance and its unique properties are essential for life. Its physical properties shape the hydrosphere and are an essential part of the water cycle and climate. Its thermodynamic properties determine evaporation and the thermal gradient in the atmosphere. The many types of precipitation involve a complex mixture of processes such as coalescence, supercooling and supersaturation. Some precipitated water becomes groundwater, and groundwater flow includes phenomena such as percolation, while the conductivity of water makes electrical and electromagnetic methods useful for tracking groundwater flow. Physical properties of water such as salinity have a large effect on its motion in the oceans.

The many phases of ice form the cryosphere and come in forms like ice sheets, glaciers, sea ice, freshwater ice, snow, and frozen ground (or permafrost).

Regions of the Earth

Size and form of the Earth

The Earth is roughly spherical, but it bulges towards the Equator, so it is roughly in the shape of an ellipsoid. This bulge is due to its rotation and is nearly consistent with an Earth in hydrostatic equilibrium. The detailed shape of the Earth, however, is also affected by the distribution of continents and ocean basins, and to some extent by the dynamics of the plates.

Structure of the Interior

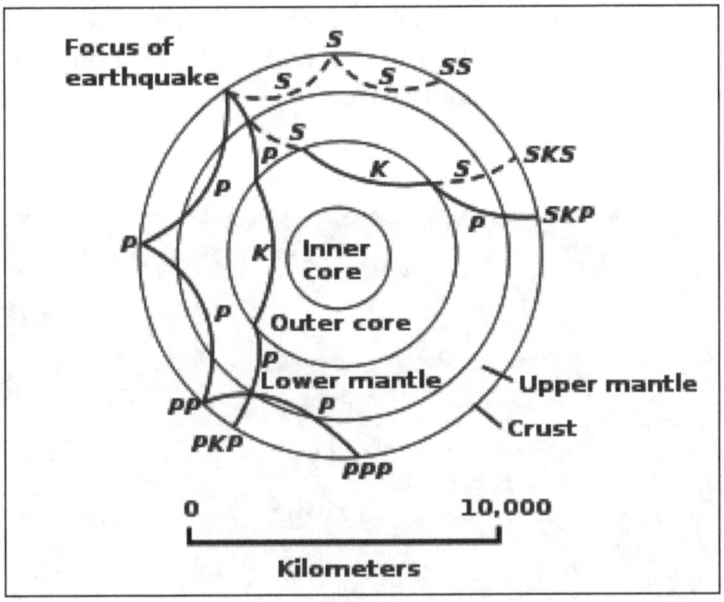

Seismic velocities and boundaries in the interior of the Earth sampled by seismic waves.

Evidence from seismology, heat flow at the surface, and mineral physics is combined with the Earth's mass and moment of inertia to infer models of the Earth's interior – its composition, density, temperature, pressure. For example, the Earth's mean specific gravity (5.515) is far higher than the typical specific gravity of rocks at the surface (2.7–3.3), implying that the deeper material is denser. This is also implied by its low moment of inertia (0.33 $M R2$, compared to 0.4 $M R2$ for a sphere of constant density). However, some of the density increase is compression under the enormous pressures inside the Earth. The effect of pressure can be calculated using the Adams–Williamson equation. The conclusion is that pressure alone cannot account for the increase in density. Instead, we know that the Earth's core is composed of an alloy of iron and other minerals.

Reconstructions of seismic waves in the deep interior of the Earth show that there are no S-waves in the outer core. This indicates that the outer core is liquid, because liquids cannot support shear. The outer core is liquid, and the motion of this highly conductive fluid generates the Earth's field. The inner core, however, is solid because of the enormous pressure.

Reconstruction of seismic reflections in the deep interior indicate some major discontinuities in seismic velocities that demarcate the major zones of the Earth: inner core, outer core, mantle, lithosphere and crust. The mantle itself is divided into the upper mantle, transition zone, lower mantle and D'' layer. Between the crust and the mantle is the Mohorovičić discontinuity.

The seismic model of the Earth does not by itself determine the composition of the layers. For a complete model of the Earth, mineral physics is needed to interpret seismic velocities in terms of composition. The mineral properties are temperature-dependent, so the geotherm must also be determined. This requires physical theory for thermal conduction and convection and the heat contribution of radioactive elements. The main model for the radial structure of the interior of the Earth is the preliminary reference Earth model (PREM). Some parts of this model have been updated by recent findings in mineral physics and supplemented by seismic tomography. The mantle is mainly composed of silicates, and the boundaries between layers of the mantle are consistent with phase transitions.

The mantle acts as a solid for seismic waves, but under high pressures and temperatures it deforms so that over millions of years it acts like a liquid. This makes plate tectonics possible. Geodynamics is the study of the fluid flow in the mantle and core.

Magnetosphere

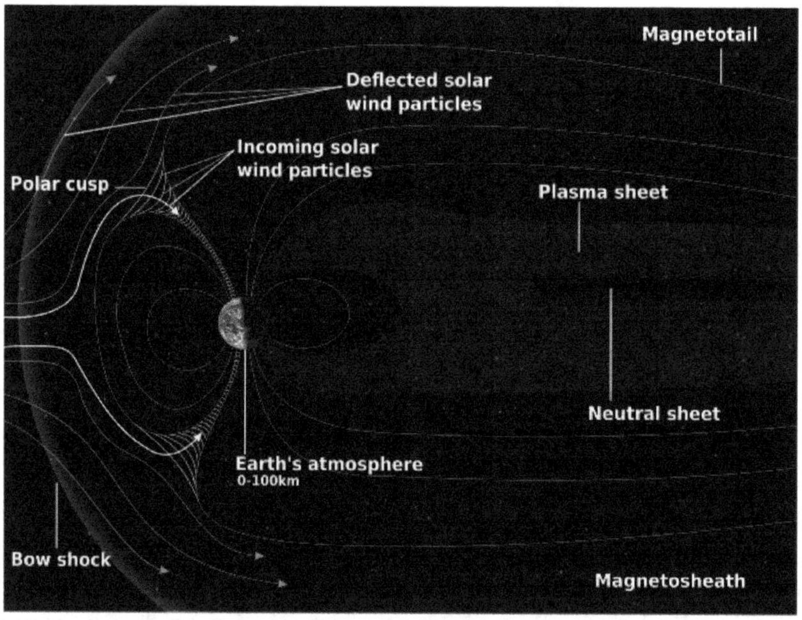

Schematic of Earth's magnetosphere. The solar wind flows from left to right.

If a planet's magnetic field is strong enough, its interaction with the solar wind forms a magnetosphere. Early space probes mapped out the gross dimensions of the Earth's magnetic field, which extends about 10 Earth radii towards the Sun. The solar wind, a stream of charged particles, streams out and around the terrestrial magnetic field, and continues behind the magnetic tail, hundreds of Earth radii downstream. Inside the magnetosphere, there are relatively dense regions of solar wind particles called the Van Allen radiation belts.

Methods

Geodesy

Geophysical measurements are generally at a particular time and place. Accurate measurements of position, along with earth deformation and gravity, are the province of geodesy. While geodesy and geophysics are separate fields, the two are so closely connected that many scientific organizations such as the American Geophysical Union, the Canadian Geophysical Union and the International Union of Geodesy and Geophysics encompass both.

Absolute positions are most frequently determined using the global positioning system (GPS). A three-dimensional position is calculated using messages from four or more visible satellites and referred to the 1980 Geodetic Reference System. An alternative, optical astronomy, combines astronomical coordinates and the local gravity vector to get geodetic coordinates. This method only provides the position in two coordinates and is more difficult to use than GPS. However, it is useful for measuring motions of the Earth such as nutation and Chandler wobble. Relative positions of two or more points can be determined using very-long-baseline interferometry.

Gravity measurements became part of geodesy because they were needed to related measurements at the surface of the Earth to the reference coordinate system. Gravity measurements on land can be made using gravimeters deployed either on the surface or in helicopter flyovers. Since the 1960s, the Earth's gravity field has been measured by analyzing the motion of satellites. Sea level can also be measured by satellites using radar altimetry, contributing to a more accurate geoid. In 2002, NASA launched the Gravity Recovery and Climate Experiment (GRACE), wherein two twin satellites map variations in Earth's gravity field by making measurements of the distance between the two satellites using GPS and a microwave ranging system. Gravity variations detected by GRACE include those caused by changes in ocean currents; runoff and ground water depletion; melting ice sheets and glaciers.

Space Probes

Space probes made it possible to collect data from not only the visible light region, but in other areas of the electromagnetic spectrum. The planets can be characterized by their force fields: gravity and their magnetic fields, which are studied through geophysics and space physics.

Measuring the changes in acceleration experienced by spacecraft as they orbit has allowed fine details of the gravity fields of the planets to be mapped. For example, in the 1970s, the gravity field disturbances above lunar maria were measured through lunar orbiters, which led to the discovery of concentrations of mass, mascons, beneath the Imbrium, Serenitatis, Crisium, Nectaris and Humorum basins.

History

Geophysics emerged as a separate discipline only in the 19th century, from the intersection of physical geography, geology, astronomy, meteorology, and physics. However, many geophysical phenomena – such as the Earth's magnetic field and earthquakes – have been investigated since the ancient era.

Ancient and Classical Eras

The magnetic compass existed in China back as far as the fourth century BC. It was used as much for feng shui as for navigation on land. It was not until good steel needles could be forged that compasses were used for navigation at sea; before that, they could not retain their magnetism long enough to be useful. The first mention of a compass in Europe was in 1190 AD.

Replica of Zhang Heng's seismoscope, possibly the first contribution to seismology.

In circa 240 BC, Eratosthenes of Cyrene deduced that the Earth was round and measured the circumference of the Earth, using trigonometry and the angle of the Sun at more than one latitude in Egypt. He developed a system of latitude and longitude.

Perhaps the earliest contribution to seismology was the invention of a seismoscope by the prolific inventor Zhang Heng in 132 AD. This instrument was designed to drop a bronze ball from the mouth of a dragon into the mouth of a toad. By looking at which of eight toads had the ball, one could determine the direction of the earthquake. It was 1571 years before the first design for a seismoscope was published in Europe, by Jean de la Hautefeuille. It was never built.

Beginnings of Modern Science

One of the publications that marked the beginning of modern science was William Gilbert's *De Magnete* (1600), a report of a series of meticulous experiments in magnetism. Gilbert deduced that compasses point north because the Earth itself is magnetic.

In 1687 Isaac Newton published his *Principia,* which not only laid the foundations for classical mechanics and gravitation but also explained a variety of geophysical phenomena such as the tides and the precession of the equinox.

The first seismometer, an instrument capable of keeping a continuous record of seismic activity, was built by James Forbes in 1844.

Geochemistry

Geochemistry is the science that uses the tools and principles of chemistry to explain the mechanisms behind major geological systems such as the Earth's crust and its oceans. The realm of geochemistry extends beyond the Earth, encompassing the entire Solar System and has made important contributions to the understanding of a number of processes including mantle convection, the formation of planets and the origins of granite and basalt.

History

The term *geochemistry* was first used by the Swiss-German chemist Christian Friedrich Schönbein in 1838. In his paper, Schönbein predicted the birth of a new field of study, stating:

"In a word, a comparative geochemistry ought to be launched, before geochemistry can become geology, and before the mystery of the genesis of our planets and their inorganic matter may be revealed."

The field began to be realised a short time after Schönbein's work, but his term - 'geochemistry' - was initially used neither by geologists nor chemists and there was much debate over which of the two sciences should be the dominant partner. There was little collaboration between geologists and chemists and the field of geochemistry remained small and unrecognised. In the late 19th Century a Swiss man by the name of Victor Goldschmidt was born, who later became known as the father of geochemistry. His paper, Geochemische Verteilungsgesetze der Elemente, on the distribution of elements in nature has been referred to as the start of geochemistry. During the early 20th Century, a number of geochemists produced work that began to popularise the field, including Frank Wigglesworth Clarke who had begun to investigate the abundances of various elements within the Earth and how the quantities were related to atomic weight. The composition of meteorites and their differences to terrestrial rocks was being investigated as early as 1850 and in 1901, Oliver C. Farrington hypothesised although there were differences, that the relative abundances should still be the same. This was the beginnings of the field of cosmochemistry and has contributed much of what we know about the formation of the Earth and the Solar System.

Subfields

Some subsets of geochemistry are:

1. Isotope geochemistry involves the determination of the relative and absolute concentrations of the elements and their isotopes in the earth and on earth's surface.
2. Examination of the distribution and movements of elements in different parts of the earth (crust, mantle, hydrosphere etc.) and in minerals with the goal to determine the underlying system of distribution and movement.
3. Cosmochemistry includes the analysis of the distribution of elements and their isotopes in the cosmos.
4. Biogeochemistry is the field of study focusing on the effect of life on the chemistry of the earth.
5. Organic geochemistry involves the study of the role of processes and compounds that are derived from living or once-living organisms.
6. Aqueous geochemistry studies the role of various elements in watersheds, including cop-

per, sulfur, mercury, and how elemental fluxes are exchanged through atmospheric-terrestrial-aquatic interactions.

7. Regional, environmental and exploration geochemistry includes applications to environmental, hydrological and mineral exploration studies.
8. Photogeochemistry is the study of light-induced chemical reactions that occur or may occur among natural components of the earth's surface.

Victor Goldschmidt is considered by most to be the father of modern geochemistry and the ideas of the subject were formed by him in a series of publications from 1922 under the title 'Geochemische Verteilungsgesetze der Elemente' (geochemical laws of distribution of the elements).

Chemical Characteristics

The more common rock constituents are nearly all oxides; chlorides, sulfides and fluorides are the only important exceptions to this and their total amount in any rock is usually much less than 1%. F. W. Clarke has calculated that a little more than 47% of the Earth's crust consists of oxygen. It occurs principally in combination as oxides, of which the chief are silica, alumina, iron oxides, and various carbonates (calcium carbonate, magnesium carbonate, sodium carbonate, and potassium carbonate). The silica functions principally as an acid, forming silicates, and all the commonest minerals of igneous rocks are of this nature. From a computation based on 1672 analyses of numerous kinds of rocks Clarke arrived at the following as the average percentage composition of the earths crust: SiO_2=59.71, Al_2O_3=15.41, Fe_2O_3=2.63, FeO=3.52, MgO=4.36, CaO=4.90, Na_2O=3.55, K_2O=2.80, H_2O=1.52, TiO_2=0.60, P_2O_5=0.22, (total 99.22%). All the other constituents occur only in very small quantities, usually much less than 1%.

These oxides combine in a haphazard way. For example, potash (potassium carbonate) and soda (sodium carbonate) combine to produce feldspars. In some cases they may take other forms, such as nepheline, leucite, and muscovite, but in the great majority of instances they are found as feldspar. Phosphoric acid with lime (calcium carbonate) forms apatite. Titanium dioxide with ferrous oxide gives rise to ilmenite. Part of the lime forms lime feldspar. Magnesium carbonate and iron oxides with silica crystallize as olivine or enstatite, or with alumina and lime form the complex ferro-magnesian silicates of which the pyroxenes, amphiboles, and biotites are the chief. Any excess of silica above what is required to neutralize the bases will separate out as quartz; excess of alumina crystallizes as corundum. These must be regarded only as general tendencies. It is possible, by rock analysis, to say approximately what minerals the rock contains, but there are numerous exceptions to any rule.

Mineral Constitution

Hence we may say that except in acid or siliceous rocks containing greater than 66% of silica are known as felsic rocks, and Quartz is not abundant. In basic rocks (containing 20% of silica or less) it is rare for them to contain as much silicon, these are referred to as mafic rocks. If Magnesium and Iron are above average while silica is low, olivine may be expected; where silica is present in greater quantity over ferro-magnesian minerals, such as augite, hornblende, enstatite or biotite, occur rather than olivine. Unless potash is high and silica relatively low, leucite will not be present, for leucite does not occur with free quartz. Nepheline, likewise, is usually found in rocks with much soda and comparatively little silica. With high alkalis, soda-bearing pyroxenes and amphiboles may be present. The lower the percentage of silica and alkali's, the greater is the prevalence of plagioclase feldspar as contracted with soda or potash feldspar. The earth crust is composed of 90% silicate minerals and their abundance in the earth is as follows; plagioclase feldspar (39%), Alkali feldspar (12%), quartz (12%), pyroxene (11%), amphiboles (5%), micas (5%), clay minerals (5%), after this the remaining silicate minerals make up another 3% of the earths crust. Only 8% of the earth is composed of non silicate minerals such as Carbonate, Oxides, and Sulfides.

The other determining factor, namely the physical conditions attending consolidation, plays on the whole a smaller part, yet is by no means negligible, as a few instances will prove. Certain minerals are practically confined to deep-seated intrusive rocks, e.g., microcline, muscovite, diallage. Leucite is very rare in plutonic masses; many minerals have special peculiarities in microscopic character according to whether they crystallized in depth or near the surface, e.g., hypersthene, orthoclase, quartz. There are some curious instances of rocks having the same chemical composition, but consisting of entirely different minerals, e.g., the hornblendite of Gran, in Norway, which contains only hornblende, has the same composition as some of the camptonites of the same locality that contain feldspar and hornblende of a different variety. In this connection we may repeat what has been said above about the corrosion of porphyritic minerals in igneous rocks. In rhyolites and trachytes, early crystals of hornblende and biotite may be found in great numbers partially converted into augite and magnetite. Hornblende and biotite were stable under the pressures and other conditions below the surface, but unstable at higher levels. In the ground-mass of these rocks, augite is almost universally present. But the plutonic representatives of the same magma, granite and syenite contain biotite and hornblende far more commonly than augite.

Felsic, Intermediate and Mafic Igneous Rocks

Those rocks that contain the most silica, and on crystallizing yield free quartz, form a group generally designated the "felsic" rocks. Those again that contain least silica and most magnesia and iron, so that quartz is absent while olivine is usually abundant, form the "mafic" group. The "intermediate" rocks include those characterized by the general absence of both quartz and olivine. An important subdivision of these contains a very high percentage of alkalis, especially soda, and consequently has minerals such as nepheline and leucite not common in other rocks. It is often separated from the others as the "alkali" or "soda" rocks, and there is a corresponding series of mafic rocks. Lastly a small sub-group rich in olivine and without feldspar has been called the "ultramafic" rocks. They have very low percentages of silica but much iron and magnesia.

Except these last, practically all rocks contain felspars or feldspathoid minerals. In the acid rocks the common feldspars are orthoclase, perthite, microcline, and oligoclase—all having much silica and alkalis. In the mafic rocks labradorite, anorthite and bytownite prevail, being rich in lime and poor in silica, potash and soda. Augite is the most common ferro-magnesian in mafic rocks, but biotite and hornblende are on the whole more frequent in felsic rocks.

Most Common Minerals	Acid	Intermediate		Mafic	Ultramafic
	Quartz Orthoclase (and Oligoclase), Mica, Hornblende, Augite	**Little or no Quartz: Orthoclase hornblende, Augite, Biotite**	**Little or no Quartz: Plagioclase Hornblende, Augite, Biotite**	**No Quartz Plagioclase Augite, Olivine**	**No Felspar Augite, Hornblende, Olivine**
Plutonic or Abyssal type	Granite	Syenite	Diorite	Gabbro	Peridotite
Intrusive or Hypabyssal type	Quartz-porphyry	Orthoclase-porphyry	Porphyrite	Dolerite	Picrite
Lavas or Effusive type	Rhyolite, Obsidian	Trachyte	Andesite	Basalt	Limburgite

Rocks that contain leucite or nepheline, either partly or a wholly replacing felspar, are not included in this table. They are essentially of intermediate or of mafic character. We might in consequence regard them as varieties of syenite, diorite, gabbro, etc., in which feldspathoid minerals occur, and indeed there are many transitions between syenites of ordinary type and nepheline — or leucite — syenite, and between gabbro or dolerite and theralite or essexite. But, as many minerals develop in these "alkali" rocks that are uncommon elsewhere, it is convenient in a purely formal classification like that outlined here to treat the whole assemblage as a distinct series.

Nepheline and Leucite-bearing Rocks			
Most Common Minerals	**Alkali Feldspar, Nepheline or Leucite, Augite, Hornblend, Biotite**	**Soda Lime Feldspar, Nepheline or Leucite, Augite, Hornblende (Olivine)**	**Nepheline or Leucite, Augite, Hornblende, Olivine**
Plutonic type	Nepheline-syenite, Leucite-syenite, Nepheline-porphyry	Essexite and Theralite	Ijolite and Missourite
Effusive type or Lavas	Phonolite, Leucitophyre	Tephrite and Basanite	Nepheline-basalt, Leucite-basalt

This classification is based essentially on the mineralogical constitution of the igneous rocks. Any chemical distinctions between the different groups, though implied, are relegated to a subordinate position. It is admittedly artificial but it has grown up with the growth of the science and is still adopted as the basis on which more minute subdivisions are erected. The subdivisions are by no means of equal value. The syenites, for example, and the peridotites, are far less important than the granites, diorites and gabbros. Moreover, the effusive andesites do not always correspond to the plutonic diorites but partly also to the gabbros. As the different kinds of rock, regarded as aggregates of minerals, pass gradually into one another, transitional types are very common and are often so important as to receive special names. The quartz-syenites and nordmarkites may be interposed between granite and syenite, the tonalites and adamellites between granite and diorite, the monzoaites between syenite and diorite, norites and hyperites between diorite and gabbro, and so on.

Geochemistry of Trace Metals in the Ocean

Trace metals readily form complexes with major ions in the ocean, including hydroxide, carbonate, and chloride and their chemical speciation changes depending on whether the environment is oxidized or reduced. Benjamin (2002) defines complexes of metals with more than one type of ligand, other than water, as mixed-ligand-complexes. In some cases, a ligand contains more than one *donor* atom, forming very strong complexes, also called chelates (the ligand is the chelator). One of the most common chelators is EDTA (ethylenediaminetetraacetic acid), which can replace six molecules of water and form strong bonds with metals that have a plus two charge. With stronger complexation, lower activity of the free metal ion is observed. One consequence of the lower reactivity of complexed metals compared to the same concentration of free metal is that the chelation tends to stabilize metals in the aqueous solution instead of in solids.

Concentrations of the trace metals cadmium, copper, molybdenum, manganese, rhenium, uranium and vanadium in sediments record the redox history of the oceans. Within aquatic environments, cadmium(II) can either be in the form CdCl+(aq) in oxic waters or CdS(s) in a reduced environment. Thus higher concentrations of Cd in marine sediments may indicate low redox potential conditions in the past. For copper(II), a prevalent form is CuCl+(aq) within oxic environments and CuS(s) and Cu2S within reduced environments. The reduced seawater environment leads to two possible oxidation states of copper, Cu(I) and Cu(II). Molybdenum is present as the Mo(VI) oxidation state as MoO42−(aq) in oxic environments. Mo(V) and Mo(IV) are present in reduced environments in the forms MoO2+(aq) and MoS2(s). Rhenium is present as the Re(VII) oxidation state as ReO4− within oxic conditions, but is reduced to Re(IV) which may form ReO2 or ReS2. Uranium is in oxidation state VI in UO2(CO3)34−(aq) and is found in the reduced form UO2(s). Vanadium is in several forms in oxidation state V(V); HVO42− and H2VO4−. Its reduced forms can include VO2+, VO(OH)3−, and V(OH)3. These relative dominance of these species depends on pH.

In the water column of the ocean or deep lakes, vertical profiles of dissolved trace metals are characterized as following *conservative–type, nutrient–type, or scavenged–type* distributions. Across these three distributions, trace metals have different residence times and are used to varying extents by planktonic microorganisms. Trace metals with conservative-type distributions have high concentrations relative to their biological use. One example of a trace metal with a conservative-type distribution is molybdenum. It has a residence time within the oceans of around 8 x 105 years and is generally present as the molybdate anion (MoO42−). Molybdenum interacts weakly with particles and displays an almost uniform vertical profile in the ocean. Relative to the abundance of molybdenum in the ocean, the amount required as a metal cofactor for enzymes in marine phytoplankton is negligible.

Trace metals with nutrient-type distributions are strongly associated with the internal cycles of particulate organic matter, especially the assimilation by plankton. The lowest dissolved concentrations of these metals are at the surface of the ocean, where they are assimilated by plankton. As dissolution and decomposition occur at greater depths, concentrations of these trace metals increase. Residence times of these metals, such as zinc, are several thousand to one hundred thousand years. Finally, an example of a scavenged-type trace metal is aluminium, which has strong interactions with particles as well as a short residence time in the ocean. The residence times of scavenged-type trace metals are around 100 to 1000 years. The concentrations of these metals are highest around bottom sediments, hydrothermal vents, and rivers. For aluminium, atmospheric dust provides the greatest source of external inputs into the ocean.

Iron and copper show hybrid distributions in the ocean. They are influenced by recycling and intense scavenging. Iron is a limiting nutrient in vast areas of the oceans, and is found in high abundance along with manganese near hydrothermal vents. Here, many iron precipitates are found, mostly in the forms of iron sulfides and oxidized iron oxyhydroxide compounds. Concentrations of iron near hydrothermal vents can be up to one million times the concentrations found in the open ocean.

Using electrochemical techniques, it is possible to show that bioactive trace metals (zinc, cobalt, cadmium, iron and copper) are bound by organic ligands in surface seawater. These ligand complexes serve to lower the bioavailability of trace metals within the ocean. For example, copper, which may be toxic to open ocean phytoplankton and bacteria, can form organic complexes. The formation of these complexes reduces the concentrations of bioavailable inorganic complexes of copper that could be toxic to sea life at high concentrations. Unlike copper, zinc toxicity in marine phytoplankton is low and there is no advantage to increasing the organic binding of Zn2+. In high nutrient-low chlorophyll regions, iron is the limiting nutrient, with the dominant species being strong organic complexes of Fe(III).

Geobiology

The colorful microbial mats of Grand Prismatic Spring in Yellowstone National Park, USA. The orange mats are composed of Chloroflexi, Cyanobacteria, and other organisms that thrive in the 70°C water. Geobiologists often study extreme environments like this because they are home to extremophilic organisms. It has been hypothesized that these environments may be representative of early Earth.

Geobiology is a field of scientific research that explores the interactions between the physical Earth and the biosphere. It is a relatively young field, and its borders are fluid. There is considerable overlap with the fields of ecology, evolutionary biology, microbiology, paleontology, and particularly biogeochemistry. Geobiology applies the principles and methods of biology and geology to the study of the ancient history of the co-evolution of life and Earth as well as the role of life in the modern world. Geobiologic studies tend to be focused on microorganisms, and on the role that life plays in altering the chemical and physical environment of the lithosphere, atmosphere, hydrosphere and/or cryosphere. It differs from biogeochemistry in that the focus is on processes and organisms over space and time rather than on global chemical cycles.

Geobiological research synthesizes the geologic record with modern biologic studies. It deals with process - how organisms affect the Earth and vice versa - as well as history - how the Earth and life have changed together. Much research is grounded in the search for fundamental understanding, but geobiology can also be applied, as in the case of microbes that clean up oil spills.

Geobiology employs molecular biology, environmental microbiology, chemical analyses, and the geologic record to investigate the evolutionary interconnectedness of life and Earth. It attempts to understand how the Earth has changed since the origin of life and what it might have been like along the way. Some definitions of geobiology even push the boundaries of this time frame - to understanding the origin of life and to the role that man has played and will continue to play in shaping the Earth in the Anthropocene.

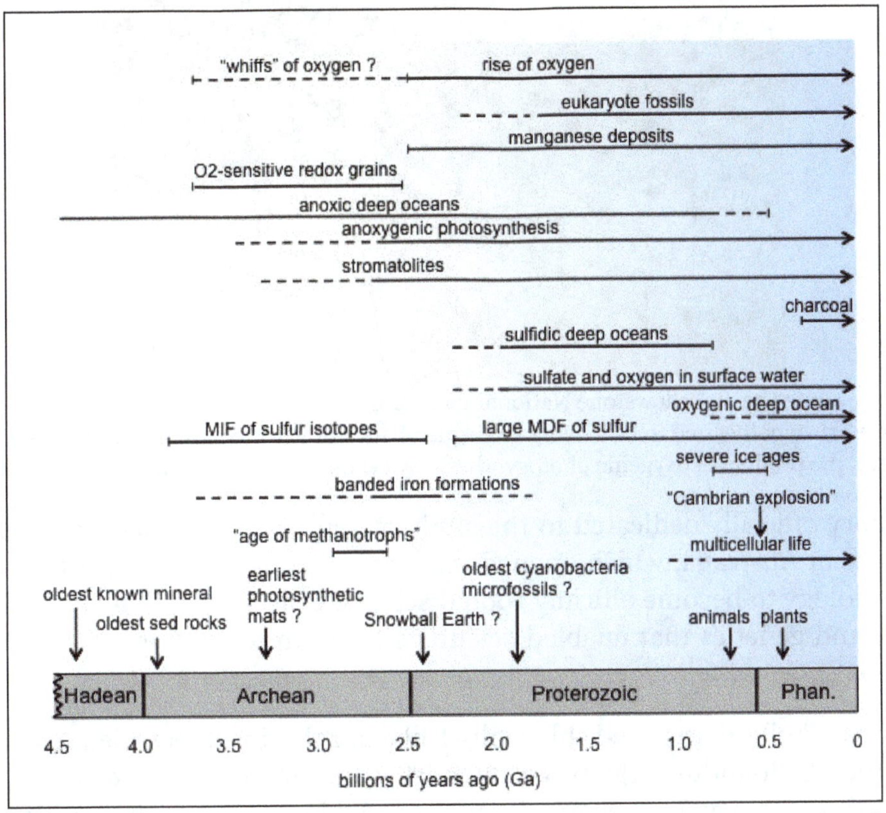

The geologic timescale overlain with major geobiologic events and occurrences. The oxygenation of the atmosphere is shown in blue starting 2.4 Ga, although the exact dating of the Great Oxygenation Event is debated.

History

The term geobiology was coined by Lourens Baas Becking in 1934. In his words, geobiology "is an attempt to describe the relationship between organisms and the Earth," for "the organism is part of the Earth and its lot is interwoven with that of the Earth." Baas Becking's definition of geobiology was born of a desire to unify environmental biology with laboratory biology. The way he practiced it aligns closely with modern environmental microbial ecology, though his definition remains applicable to all of geobiology. In his book, Geobiology, Bass Becking stated that he had no intention of inventing a new field of study.

Baas Becking's understanding of geobiology was heavily influenced by his predecessors, including

Martinus Beyerinck, his teacher from the Dutch School of Microbiology. Others included Vladimir Vernadsky, who argued that life changes the surface environment of Earth in The Biosphere, his 1926 book, and Sergei Vinogradsky, famous for discovering lithotrophic bacteria.

A microbial mat in White Creek, Yellowstone National Park, USA. Note the conical microstructure of the bacterial communities. These are hypothesized to be a living analogue of ancient fossil stromatolites. Each cone has an oxygen gas bubble on top, the product of oxygenic photosynthesis by cyanobacteria in the multi-species microbial mats.

The first laboratory officially dedicated to the study of geobiology was the Baas Becking Geobiological Laboratory in Australia, which opened its doors in 1965. However, it took another 40 or so years for geobiology to become a firmly rooted scientific discipline, thanks in part to advances in geochemistry and genetics that enabled scientists to begin to synthesize the study of life and planet.

In the 1930s, Alfred Treibs discovered chlorophyll-like porphyrins in petroleum, confirming its biological origin, thereby founding organic geochemistry and establishing the notion of biomarkers, a critical aspect of geobiology. But several decades passed before the tools were available to begin to search in earnest for chemical marks of life in the rocks. In the 1970s and 80s, scientists like Geoffrey Eglington and Roger Summons began to find lipid biomarkers in the rock record using equipment like GCMS.

On the biology side of things, in 1977, Carl Woese and George Fox published a phylogeny of life on Earth, including a new domain - the Archaea. And in the 1990s, genetics and genomics studies became possible, broadening the scope of investigation of the interaction of life and planet.

Today, geobiology has its own journals, such as *Geobiology, established in 2003,* and *Biogeosciences, established in 2004,* as well as recognition at major scientific conferences. It got its own Gordon Research Conference in 2011, a number of geobiology textbooks have been published, and many universities around the world offer degree programs in geobiology.

Major Geobiological Events

Graphical Anthropocentric representation of Earth's history as a spiral.

Perhaps the most profound geobiological event is the introduction of oxygen into the atmosphere by photosynthetic bacteria. This oxygenation of Earth's primoidial atmosphere (the so-called oxygen catastrophe or Great Oxygenation Event) and the oxygenation of the oceans altered surface biogeochemical cycles and the types of organisms that have been evolutionarily selected for.

A subsequent major change was the advent of multicellularity. The presence of oxygen allowed eukaryotes and, later, multicellular life to evolve.

More anthropocentric geobiologic events include the origin of animals and the establishment of terrestrial plant life, which affected continental erosion and nutrient cycling, and likely changed the types of rivers observed, allowing channelization of what were previously predominantly braided rivers.

More subtle geobiological events include the role of termites in overturning sediments, coral reefs in depositing calcium carbonate and breaking waves, sponges in absorbing dissolved marine silica, the role of dinosaurs in breaching river levees and promoting flooding, and the role of large mammal dung in distributing nutrients.

Important Concepts

Geobiology is founded upon a few core concepts that unite the study of Earth and life. While there are many aspects of studying past and present interactions between life and Earth that are unclear, several important ideas and concepts provide a basis of knowledge in geobiology that serve as a platform for posing researchable questions, including the evolution of life and planet and the co-evolution of the two, genetics - from both a historical and functional standpoint, the metabolic diversity of all life, the sedimentological preservation of past life, and the origin of life.

Co-Evolution of Life and Earth

A core concept in geobiology is that life changes over time through evolution. The theory of evolution postulates that unique populations of organisms or species arose from genetic modifications in the ancestral population which were passed down by drift and natural selection.

Along with standard biological evolution, life and planet co-evolve. Since the best adaptations are those that suit the ecological niche that the organism lives in, the physical and chemical characteristics of the environment drive the evolution of life by natural selection, but the opposite can also be true: with every advent of evolution, the environment changes.

A classic example of co-evolution is the evolution of oxygen-producing photosynthetic cyanobacteria which oxygenated Earth's Archean atmosphere. The ancestors of cyanobacteria began using water as an electron source to harness the energy of the sun and expelling oxygen before or during the early Paleoproterozoic. During this time, around 2.4 to 2.1 billion years ago, geologic data suggests that atmospheric oxygen began to rise in what is termed the Great Oxygenation Event (GOE). It is unclear for how long cyanobacteria had been doing oxygenic photosynthesis before the GOE. Some evidence suggests there were geochemical "buffers" or sinks suppressing the rise of oxygen such as volcanism though cyanobacteria may have been around producing it before the GOE. Other evidence indicates that the rise of oxygenic photosynthesis was coincident with the GOE.

Banded iron formation (BIF), Hammersley Formation, Western Australia.

The presence of oxygen on Earth from its first production by cyanobacteria to the GOE and through today has drastically impacted the course of evolution of life and planet. It may have triggered the formation of oxidized minerals and the disappearance of oxidizable minerals like pyrite from ancient stream beds. The presence of banded-iron formations (BIFs) have been interpreted as a clue for the rise of oxygen since small amounts of oxygen could have reacted with reduced ferrous iron (Fe(II)) in the oceans, resulting in the deposition of sediments containing Fe(III) oxide in places like Western Australia. However, any oxidizing environment, including that provided by microbes such as the iron-oxidizing photoautotroph *Rhodopseudomonas palustris,* can trigger iron oxide formation and thus BIF deposition. Other mechanisms include oxidation by UV light. Indeed, BIFs occur across large swaths of Earth's history and may not correlate with only one event.

Other changes correlated with the rise of oxygen include the appearance of rust-red ancient paleosols, different isotope fractionation of elements such as sulfur, and global glaciations and Snowball Earth events, perhaps caused by the oxidation of methane by oxygen, not to mention an overhaul of the types of organisms and metabolisms on Earth. Whereas organisms prior to the rise of oxygen were likely poisoned by oxygen gas as many anaerobes are today, those that evolved ways to harness the electron-accepting and energy-giving power of oxygen were poised to thrive and colonize the aerobic environment.

Modern, living stromatolites in Shark Bay, Australia. Shark Bay is one of the few places in the world where stromatolites can be seen today, though they were likely common in ancient shallow seas before the rise of metazoan predators.

The Earth has Changed

Earth has not remained the same since its planetary formation 4.5 billion years ago. Continents have formed, broken up, and collided, offering new opportunities for and barriers to the dispersal of life. The redox state of the atmosphere and the oceans has changed, as indicated by isotope data. Fluctuating quantities of inorganic compounds such as carbon dioxide, nitrogen, methane, and oxygen have been driven by life evolving new biological metabolisms to make these chemicals and have driven the evolution of new metabolisms to use those chemicals. Earth acquired a magnetic field about 3.4 Ga that has undergone a series of geomagnetic reversals on the order of millions of years. The surface temperature is in constant fluctuation, falling in glaciations and Snowball Earth events due to ice-albedo feedback, rising and melting due to volcanic outgassing, and stabilizing due to silicate weathering feedback.

And the Earth is not the only one that changed - the luminosity of the sun has increased over time. Because rocks record a history of relatively constant temperatures since Earth's beginnings, there must have been more greenhouse gasses to keep the temperatures up in the Archean when the sun was younger and fainter. All these major differences in the environment of the Earth placed very different constraints on the evolution of life throughout our planet's history. Moreover, more subtle changes in the habitat of life are always occurring, shaping the organisms and traces that we observe today and in the rock record.

Genes Encode Geobiological Function and History

The genetic code is key to observing the history of evolution and understanding the capabilities of organisms. Genes are the basic unit of inheritance and function and, as such, they are the basic unit of evolution and the means behind metabolism.

Phylogeny Predicts Evolutionary History

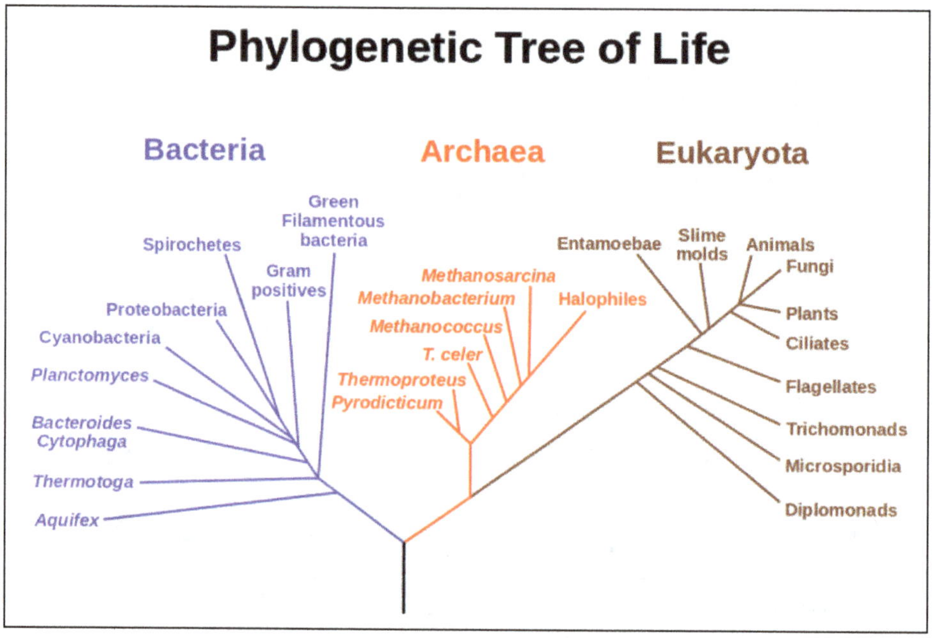

A phylogenetic tree of living things, based on rRNA data and proposed by Carl Woese, showing the separation of bacteria, archaea, and eukaryotes and linking the three branches of living organisms to the LUCA (the black trunk at the bottom of the tree).

Phylogeny takes genetic sequences from living organisms and compares them to each other to reveal evolutionary relationships, much like a family tree reveals how individuals are connected to their distant cousins. It allows us to decipher modern relationships and infer how evolution happened in the past.

Phylogeny can give some sense of history when combined with a little bit more information. Each difference in the DNA indicates divergence between one species and another. This divergence, whether via drift or natural selection, is representative of some lapse of time. Comparing DNA sequences alone gives a record of the history of evolution with an arbitrary measure of phylogenetic distance "dating" that last common ancestor. However, if information about the rate of genetic mutation is available or geologic markers are present to calibrate evolutionary divergence (i.e. fossils), we have a timeline of evolution. From there, with an idea about other contemporaneous changes in life and environment, we can begin to speculate why certain evolutionary paths might have been selected for.

Genes Encode Metabolism

Molecular biology allows scientists to understand a gene's function using microbial culturing and mutagenesis. Searching for similar genes in other organisms and in metagenomic and metatranscriptomic data allows us to understand what processes could be relevant and important in a given ecosystem, providing insight into the biogeochemical cycles in that environment.

For example, an intriguing problem in geobiology is the role of organisms in the global cycling of methane. Genetics has revealed that the methane monooxygenase gene (*pmo)* is used for oxidizing methane and is present in all aerobic methane-oxidizers, or methanotrophs. The presence of DNA sequences of the *pmo* gene in the environment can be used as a proxy for methanotrophy. A more generalizable tool is the 16S ribosomal RNA gene, which is found in bacteria and archaea. This gene evolves very slowly over time and is not usually horizontally transferred, and so it is often used to distinguish different taxonomic units of organisms in the environment. In this way, genes are clues to organismal metabolism and identity. Genetics enables us to ask 'who is there?' and 'what are they doing?' This approach is called metagenomics.

3.4 billion year-old stromatolites from the Warrawoona Group, Western Australia. While the origin of Precambrian stromatolites is a heavily debated topic in geobiology, stromatolites from Warrawoona are hypothesized to have been formed by ancient communities of microbes.

Metabolic Diversity Influences the Environment

Life harnesses chemical reactions to generate energy, perform biosynthesis, and eliminate waste. Different organisms use very different metabolic approaches to meet these basic needs. While animals such as ourselves are limited to aerobic respiration, other organisms can "breathe" sulfate (SO42-), nitrate (NO3-), ferric iron (Fe(III)), and uranium (U(VI)), or live off energy from fermentation. Some organisms, like plants, are autotrophs, meaning that they can fix carbon dioxide for biosynthesis. Plants are photoautotrophs, in that they use the energy of light to fix carbon. Microorganisms employ oxygenic and anoxygenic photoautotrophy, as well as chemoautotrophy. Microbial communities can coordinate in syntrophic metabolisms to shift reaction kinetics in their favor. Many organisms can perform multiple metabolisms to achieve the same end goal; these are called mixotrophs.

Biotic metabolism is directly tied to the global cycling of elements and compounds on Earth. The geochemical environment fuels life, which then produces different molecules that go into the external environment. (This is directly relevant to biogeochemistry.) In addition, biochemical reactions are catalyzed by enzymes which sometimes prefer one isotope over others. For example, oxygenic photosynthesis is catalyzed by RuBisCO, which prefers carbon-12 over carbon-13, resulting in carbon isotope fractionation in the rock record.

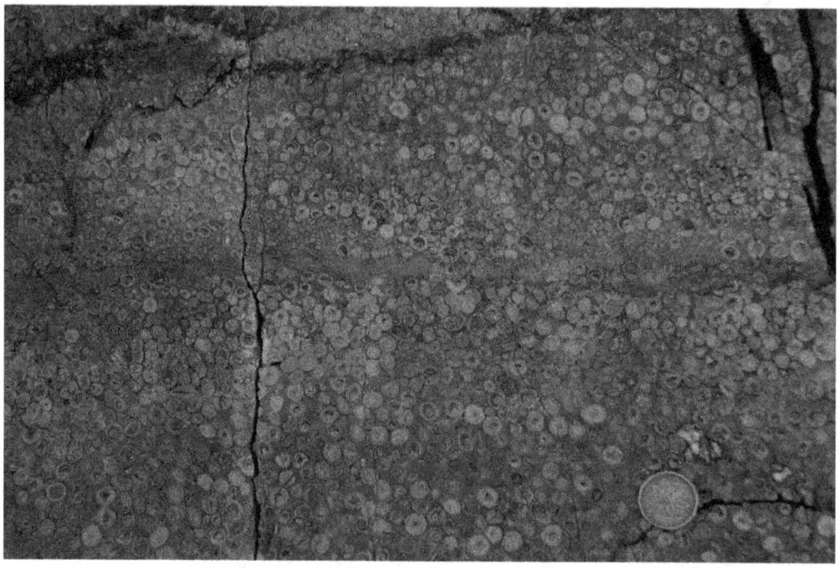

"Giant" ooids of the Johnnie Formation in the Death Valley area, California, USA. Ooids are near-spheroidal calcium carbonate grains that accumulate around a central nucleus and can be sedimented to form oolite like this. Microbes can mediate the formation of ooids.

Sedimentary Rocks Tell a Story

Sedimentary rocks preserve remnants of the history of life on Earth in the form of fossils, biomarkers, isotopes, and other traces. The rock record is far from perfect, and the preservation of biosignatures is a rare occurrence. Understanding what factors determine the extent of preservation and the meaning behind what is preserved are important components to detangling the ancient history of the co-evolution of life and Earth. The sedimentary record allows scientists to observe changes in life and Earth in composition over time and sometimes even date major transitions, like extinction events.

Some classic examples of geobiology in the sedimentary record include stromatolites and banded-iron formations. The role of life in the origin of both of these is a heavily debated topic.

Life is Fundamentally Chemistry

The first life arose from abiotic chemical reactions. When this happened, how it happened, and even what planet it happened on are uncertain. However, life follows the rules of and arose from lifeless chemistry and physics. It is constrained by principles such as thermodynamics. This is an important concept in the field because it is represents the epitome of the interconnectedness, if not sameness, of life and Earth.

While often delegated to the field of astrobiology, attempts to understand how and when life arose are relevant to geobiology as well. The first major strides towards understanding the "how" came with the Miller-Urey experiment, when amino acids formed out of a simulated "primordial soup". Another theory is that life originated in a system much like the hydrothermal vents at mid-oceanic spreading centers. In the Fischer-Tropsch synthesis, a variety of hydrocarbons form under vent-like conditions. Other ideas include the "RNA World" hypothesis, which postulates that the first biologic molecule was RNA and the idea that life originated elsewhere in the solar system and was brought to Earth, perhaps via a meteorite.

Methodology

A microbial mat growing on acidic soil in Norris Geyser basin, Yellowstone National Park, USA. The black top serves as a sort of sunscreen, and when you look underneath you see the green cyanobacteria.

While geobiology is a diverse and varied field, encompassing ideas and techniques from a wide range of disciplines, there are a number of important methods that are key to the study of the interaction of life and Earth that are highlighted here.

1. *Laboratory culturing of microbes is used to characterize the metabolism and lifestyle of organisms of interest.*

2. *Gene sequencing allows scientists to study the relationships between extant organisms using phylogenetics.*

3. *Experimental genetic manipulation or* mutagenesis is used to determine the function of genes in living organisms.

4. *Microscopy is used to visualize the microbial world. Microscope work ranges from environmental observation to quantitative studies with* DNA probes to high-definition visualization of the microbe-mineral interface by electron microscope (EM).

5. *Isotope tracers can be used to track biochemical reactions to understand microbial metabolism.*
6. *Isotope natural abundance in rocks can be measured to look for* isotopic fractionation that is consistent with biologic origin.
7. Detailed *environmental characterization is important to understanding what about a habitat might be driving life's evolution and, in turn, how life might be changing that niche. It includes and is not limited to, temperature, light, pH, salinity, concentration of specific molecules like oxygen, and the biologic community.*
8. *Sedimentology and stratigraphy are used to read the rocks. The rock record stores a history of geobiologic processes in sediments which can be unearthed through an understanding of* deposition, sedimentation, compaction, diagenesis, and deformation.
9. The search for and study of fossils, while often delegated to the separate field of *paleontology, is important in geobiology, though the scale of fossils is typically smaller* (micropaleontology).
10. The biochemical analysis of *biomarkers, which are fossilized or modern molecules that are indicative of the presence of a certain group of organisms or metabolism, is used to answer the evidence for life and metabolic diversity questions.*
11. *Paleomagnetics is the study of the planet's ancient magnetic field. It is significant to understanding* magnetofossils, biomineralization, and global ecosystem changes.

Sub-Disciplines and Related Fields

As its name suggests, geobiology is closely related to many other fields of study, and does not have clearly defined boundaries or perfect agreement on what exactly they comprise. Some practitioners take a very broad view of its boundaries, encompassing many older, more established fields such as biogeochemistry, paleontology, and microbial ecology. Others take a more narrow view, assigning it to emerging research that falls between these existing fields, such as with geomicrobiology. The following list includes both those that are clearly a part of geobiology, e.g. geomicrobiology, as well as those that share scientific interests but have not historically been considered a sub-discipline of geobiology, e.g. paleontology.

Astrobiology

Astrobiology is an interdisciplinary field that uses a combination of geobiological and planetary science data to establish a context for the search for life on other planets. The origin of life from non-living chemistry and geology, or abiogenesis, is a major topic in astrobiology. Even though it is fundamentally an earth-bound concern, and therefore of great geobiological interest, getting at the origin of life necessitates considering what life requires, what, if anything, is special about Earth, what might have changed to allow life to blossom, what constitutes evidence for life, and even what constitutes life itself. These are the same questions that scientists might ask when searching for alien life. In addition, astrobiologists research the possibility of life based on other metabolisms and elements, the survivability of Earth's organisms on other planets or spacecrafts, planetary and solar system evolution, and space geochemistry.

Biogeochemistry

Biogeochemistry is a systems science that synthesizes the study of biological, geological, and chemical processes to understand the reactions and composition of the natural environment. It is concerned primarily with global elemental cycles, such as that of nitrogen and carbon. The father of biogeochemistry was James Lovelock, whose "Gaia hypothesis" proposed that Earth's biological, chemical, and geologic systems interact to stabilize the conditions on Earth that support life.

Geobiochemistry

Stromatolites in the Green River Shale, Wyoming, USA, dating to the Eocene.

Geobiochemistry is similar to biogeochemistry, but differs by placing emphasis on the effects of geology on the development of life's biochemical processes, as distinct from the role of life on Earth's cycles. Its primary goal is to link biological changes, encompassing evolutionary modifications of genes and changes in the expression of genes and proteins, to changes in the temperature, pressure, and composition of geochemical processes to understand when and how metabolism evolved. Geobiochemistry is founded on the notion that life is a planetary response because metabolic catalysis enables the release of energy trapped by a cooling planet.

Environmental Microbiology

Microbiology is a broad scientific discipline pertaining to the study of that life which is best viewed under a microscope. It encompasses several fields that are of direct relevance to geobiology, and the tools of microbiology all pertain to geobiology. Environmental microbiology is especially entangled in geobiology since it seeks an understanding of the actual organisms and processes that are relevant in nature, as opposed to the traditional lab-based approach to microbiology. Microbial ecology is similar, but tend to focus more on lab studies and the relationships between organisms within a community, as well as within the ecosystem of their chemical and geological physical environment. Both rely on techniques such as sample collection from diverse environments, metagenomics, DNA sequencing, and statistics.

Geomicrobiology and Microbial Geochemistry

A vertical cross section of a microbial mat containing different organisms that perform different metabolisms. The green are presumably cyanobacteria, and teepee-like microstructures are visible on the surface.

Geomicrobiology traditionally studies the interactions between microbes and minerals. While it is generally reliant on the tools of microbiology, microbial geochemistry uses geological and chemical methods to approach the same topic from the perspective of the rocks. Geomicrobiology and microbial geochemistry (GMG) is a relatively new interdisciplinary field that more broadly takes on the relationship between microbes, Earth, and environmental systems. Billed as a subset of both geobiology and geochemistry, GMG seeks to understand elemental biogeochemical cycles and the evolution of life on Earth. Specifically, it asks questions about where microbes live, their local and global abundance, their structural and functional biochemistry, how they have evolved, biomineralization, and their preservation potential and presence in the rock record. In many ways, GMG appears to be equivalent to geobiology, but differs in scope: geobiology focuses on the role of all life, while GMG is strictly microbial. Regardless, it is these tiniest creatures that dominated to history of life integrated over time and seem to have had the most far-reaching effects.

Molecular Geomicrobiology

Molecular geomicrobiology takes a mechanistic approach to understanding biological processes that are geologically relevant. It can be at the level of DNA, protein, lipids, or any metabolite.

Organic Geochemistry

Organic geochemistry is the study of organic molecules that appear in the fossil record in sedimentary rocks. Research in this field concerns molecular fossils that are often lipid biomarkers. Molecules like sterols and hopanoids, membrane lipids found in eukaryotes and bacteria, respectively, can be preserved in the rock record on billion-year timescales. Following the death of the organism they came from and sedimentation, they undergo a process called diagenesis whereby many of the specific functional groups from the lipids are lost, but the hydrocarbon skeleton remains intact. These fossilized lipids are called steranes and hopanes, respectively. There are also other types of molecular fossils, like porphyrins, the discovery of which in petroleum by Alfred E. Treibs actually led to the invention of the field. Other aspects of geochemistry that are also pertinent to geobiology include isotope geochemistry, in which scientists search for isotope fractionation in the rock record, and the chemical analysis of biominerals, such as magnetite or microbially-precipitated gold.

Ediacaran fossils from Mistaken Point, Newfoundland. Ediacaran biota originated during the Ediacaran Period and are unlike most animals around today.

Paleontology

Perhaps the oldest of the bunch, paleontology is the study of fossils. It involves the discovery, excavation, dating, and paleoecological understanding of any type of fossil, microbial or dinosaur, trace or body fossil. Micropaleontology is particularly relevant to geobiology. Putative bacterial microfossils and ancient stromatolites are used as evidence for the rise of metabolisms such as oxygenic photosynthesis. The search for molecular fossils, such as lipid biomarkers like steranes and hopanes, has also played an important role in geobiology and organic geochemistry. Relevant sub-disciples include paleoecology and paleobiogeoraphy.

Biogeography

Biogeography is the study of the geographic distribution of life through time. It can look at the present distribution of organisms across continents or between microniches, or the distribution of organisms through time, or in the past, which is called paleobiogeography.

Evolutionary Biology

Evolutionary biology is the study of the evolutionary processes that have shaped the diversity of life on Earth. It incorporates genetics, ecology, biogeography, and paleontology to analyze topics including natural selection, variance, adaptation, divergence, genetic drift, and speciation.

Geoinformatics

Geoinformatics is the science and the technology which develops and uses information science infrastructure to address the problems of geography, cartography, geosciences and related branches of science and engineering.

Overview

Geoinformatics has been described as "the science and technology dealing with the structure and character of spatial information, its capture, its classification and qualification, its storage, processing, portrayal and dissemination, including the infrastructure necessary to secure optimal use of this information" or "the art, science or technology dealing with the acquisition, storage, processing production, presentation and dissemination of geoinformation".

Geomatics is a similarly used term which encompasses geoinformatics, but geomatics focuses more so on surveying. Geoinformatics has at its core the technologies supporting the processes of acquiring, analyzing and visualizing spatial data. Both geomatics and geoinformatics include and rely heavily upon the theory and practical implications of geodesy.

Geography and earth science increasingly rely on digital spatial data acquired from remotely sensed images analyzed by geographical information systems (GIS) and visualized on paper or the computer screen.

Geoinformatics combines geospatial analysis and modeling, development of geospatial databases, information systems design, human-computer interaction and both wired and wireless networking technologies. Geoinformatics uses geocomputation and geovisualization for analyzing geoinformation.

Branches of geoinformatics include:

Cartography.

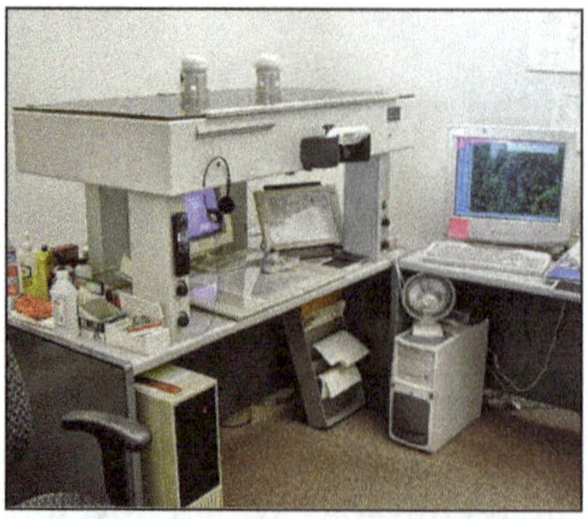

Photogrammetry.

Remote Sensing.

Geoinformatics Research

Research in this field is used to support global and local environmental, energy and security programs. The Geographic Information Science and Technology group of Oak Ridge National Laboratory is supported by various government departments and agencies including the United States Department of Energy. It is currently the only group in the United States Department of Energy National Laboratory System to focus on advanced theory and application research in this field. There are also a lot of interdiscipline research involved in geoinformatics fields including computer science, information technology, software engineering, biogeography, geography, conservation, architecture, spatial analysis and reinformacement learning.

Applications

Many fields benefit from geoinformatics, including urban planning and land use management, in-car navigation systems, virtual globes, public health, local and national gazetteer management, environmental modeling and analysis, military, transport network planning and management, agriculture, meteorology and climate change, oceanography and coupled ocean and atmosphere modelling, business location planning, architecture and archeological reconstruction, telecommunications, criminology and crime simulation, aviation, biodiversity conservation and maritime transport. The importance of the spatial dimension in assessing, monitoring and modelling various issues and problems related to sustainable management of natural resources is recognized all over the world. Geoinformatics becomes very important technology to decision-makers across a wide range of disciplines, industries, commercial sector, environmental agencies, local and national government, research, and academia, national survey and mapping organisations, International organisations, United Nations, emergency services, public health and epidemiology, crime mapping, transportation and infrastructure, information technology industries, GIS consulting firms, environmental management agencies), tourist industry, utility companies, market analysis and e-commerce, mineral exploration, etc. Many government and non government agencies started to use spatial data for managing their day-to-day activities.

Atmospheric Chemistry

Atmospheric chemistry is a branch of atmospheric science in which the chemistry of the Earth's atmosphere and that of other planets is studied. It is a multidisciplinary approach of research and draws on environmental chemistry, physics, meteorology, computer modeling, oceanography, geology and volcanology and other disciplines. Research is increasingly connected with other arenas of study such as climatology.

The composition and chemistry of the Earth's atmosphere is of importance for several reasons, but primarily because of the interactions between the atmosphere and living organisms. The composition of the Earth's atmosphere changes as result of natural processes such as volcano emissions, lightning and bombardment by solar particles from corona. It has also been changed by human activity and some of these changes are harmful to human health, crops and ecosystems. Examples of problems which have been addressed by atmospheric chemistry include acid rain, ozone depletion, photochemical smog, greenhouse gases and global warming. Atmospheric chemists seek to understand the causes of these problems, and by obtaining a theoretical understanding of them, allow possible solutions to be tested and the effects of changes in government policy evaluated.

Atmospheric Composition

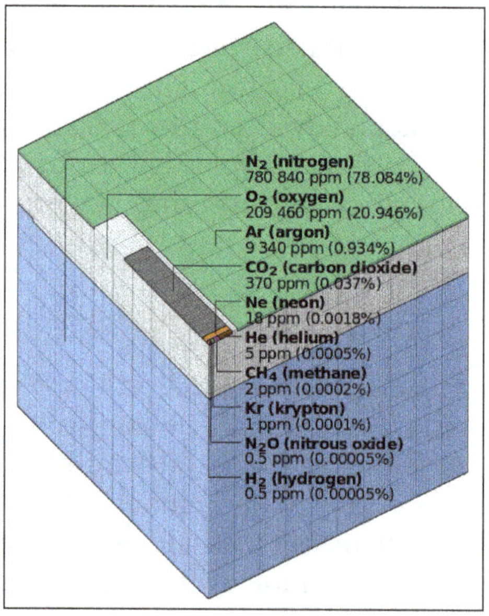

Visualisation of composition by volume of Earth's atmosphere. Water vapour is not included as it is highly variable. Each tiny cube (such as the one representing krypton) has one millionth of the volume of the entire block. Data is from NASA Langley.

Average composition of dry atmosphere (mole fractions)		
Gas	**per NASA**	
Nitrogen, N_2	78.084%	
Oxygen, O_2	20.946%	
Minor constituents (mole fractions in ppm)		
Argon, Ar	9340	
Carbon dioxide, CO_2	400	

Neon, Ne	18.18	
Helium, He	5.24	
Methane, CH_4	1.7	
Krypton, Kr	1.14	
Hydrogen, H_2	0.55	
Nitrous oxide, N_2O	0.5	
Xenon, Xe	0.09	
Nitrogen dioxide, NO_2	0.02	
Water		
Water vapour	Highly variable; typically makes up about 1%	

Notes: the concentration of CO2 and CH4 vary by season and location. The mean molecular mass of air is 28.97 g/mol. Ozone (O3) is not included due to its high variability.

History

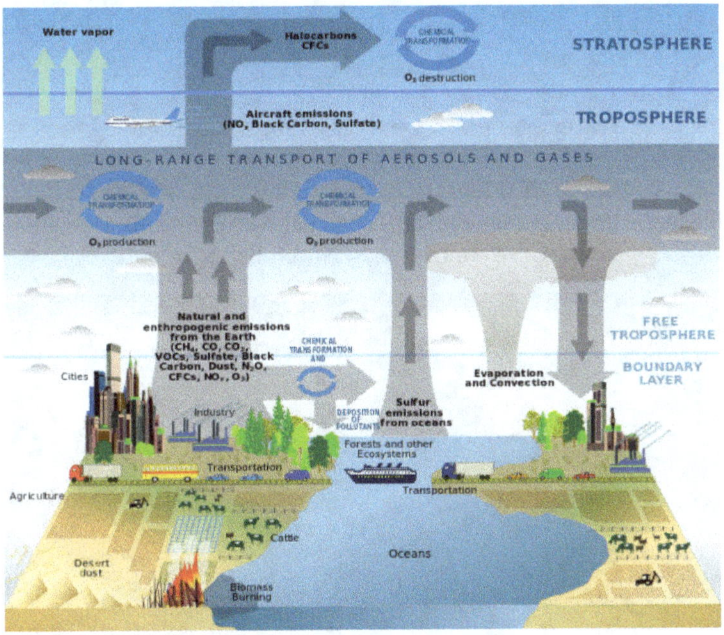

Schematic of chemical and transport processes related to atmospheric composition.

The ancient Greeks regarded air as one of the four elements, but the first scientific studies of atmospheric composition began in the 18th century. Chemists such as Joseph Priestley, Antoine Lavoisier and Henry Cavendish made the first measurements of the composition of the atmosphere.

In the late 19th and early 20th centuries interest shifted towards trace constituents with very small concentrations. One particularly important discovery for atmospheric chemistry was the discovery of ozone by Christian Friedrich Schönbein in 1840.

In the 20th century atmospheric science moved on from studying the composition of air to a consideration of how the concentrations of trace gases in the atmosphere have changed over time and the chemical processes which create and destroy compounds in the air. Two particularly important examples of this were the explanation by Sydney Chapman and Gordon Dobson of how the ozone layer is created and maintained, and the explanation of photochemical smog by Arie Jan Haagen-Smit. Further studies on ozone issues led to the 1995 Nobel Prize in Chemistry award shared between Paul Crutzen, Mario Molina and Frank Sherwood Rowland.

In the 21st century the focus is now shifting again. Atmospheric chemistry is increasingly studied as one part of the Earth system. Instead of concentrating on atmospheric chemistry in isolation the focus is now on seeing it as one part of a single system with the rest of the atmosphere, biosphere and geosphere. An especially important driver for this is the links between chemistry and climate such as the effects of changing climate on the recovery of the ozone hole and vice versa but also interaction of the composition of the atmosphere with the oceans and terrestrial ecosystems.

Carbon dioxide in Earth's atmosphere if *half of* global-warming emissions are *not absorbed.* (NASA simulation; 9 November 2015)

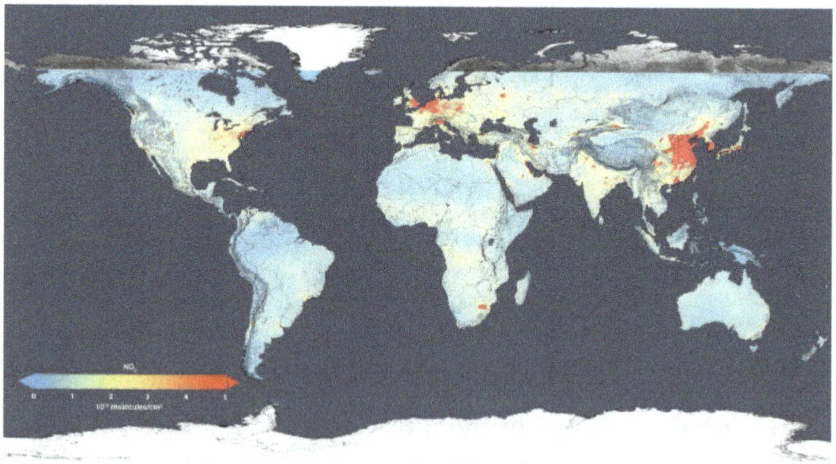

Nitrogen dioxide 2014 - global air quality levels (released 14 December 2015).

Methodology

Observations, lab measurements and modeling are the three central elements in atmospheric chemistry. Progress in atmospheric chemistry is often driven by the interactions between these components and they form an integrated whole. For example, observations may tell us that more of a chemical compound exists than previously thought possible. This will stimulate new modelling and laboratory studies which will increase our scientific understanding to a point where the observations can be explained.

Observation

Observations of atmospheric chemistry are essential to our understanding. Routine observations of chemical composition tell us about changes in atmospheric composition over time. One important example of this is the Keeling Curve - a series of measurements from 1958 to today which show a steady rise in of the concentration of carbon dioxide. Observations of atmospheric chemistry are made in observatories such as that on Mauna Loa and on mobile platforms such as aircraft (e.g. the UK's Facility for Airborne At-mospheric Measurements), ships and balloons. Observations of atmospheric composition are in-creasingly made by satellites with important instruments such as GOME and MOPITT giving a global picture of air pollution and chemistry. Surface observations have the advantage that they provide long term records at high time resolution but are limited in the vertical and horizontal space they provide observations from. Some surface based instruments e.g. LIDAR can provide concentration profiles of chemical compounds and aerosol but are still restricted in the horizontal region they can cover. Many observations are available on line in Atmospheric Chemistry Observational Databases.

Lab Measurements

Measurements made in the laboratory are essential to our understanding of the sources and sinks of pollutants and naturally occurring compounds. Lab studies tell us which gases react with each other and how fast they react. Measurements of interest include reactions in the gas phase, on surfaces and in water. Also of high importance is photochemistry which quantifies how quickly molecules are split apart by sunlight and what the products are plus thermodynamic data such as Henry's law coefficients.

Modeling

In order to synthesise and test theoretical understanding of atmospheric chemistry, computer models (such as chemical transport models) are used. Numerical models solve the differential equations governing the concentrations of chemicals in the atmosphere. They can be very simple or very complicated. One common trade off in numerical models is between the number of chemical compounds and chemical reactions modelled versus the representation of transport and mixing in the atmosphere. For example, a box model might include hundreds or even thousands of chemical reactions but will only have a very crude representation of mixing in the atmosphere. In contrast, 3D models represent many of the physical processes of the atmosphere but due to constraints on computer resources will have far fewer chemical reactions and compounds. Models can be used to interpret observations, test understanding of chemical reactions and predict future concentrations of chemical compounds in the atmosphere. One important current trend is for atmospheric chemistry modules to become one part of earth system models in which the links between climate, atmospheric composition and the biosphere can be studied.

Some models are constructed by automatic code generators (e.g. Autochem or KPP). In this approach a set of constituents are chosen and the automatic code generator will then select the reactions involving those constituents from a set of reaction databases. Once the reactions have been chosen the ordinary differential equations (ODE) that describe their time evolution can be automatically constructed.

Atmospheric physics

Atmospheric physics is the application of physics to the study of the atmosphere. Atmospheric physicists attempt to model Earth's atmosphere and the atmospheres of the other planets using fluid flow equations, chemical models, radiation budget, and energy transfer processes in the atmosphere (as well as how these tie into other systems such as the oceans). In order to model weather systems, atmospheric physicists employ elements of scattering theory, wave propagation models, cloud physics, statistical mechanics and spatial statistics which are highly mathematical and related to physics. It has close links to meteorology and climatology and also covers the design and construction of instruments for studying the atmosphere and the interpretation of the data they provide, including remote sensing instruments. At the dawn of the space age and the introduction of sounding rockets, aeronomy became a subdiscipline concerning the upper layers of the atmosphere, where dissociation and ionization are important.

Remote Sensing

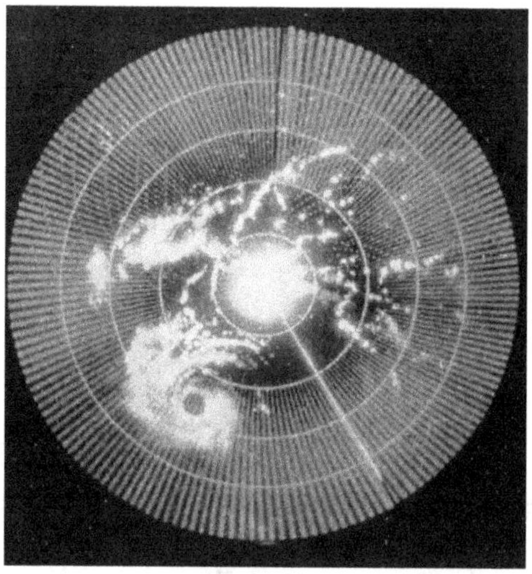

Brightness can indicate reflectivity as in this 1960 weather radar image (of Hurricane Abby). The radar's frequency, pulse form, and antenna largely determine what it can observe.

Remote sensing is the small or large-scale acquisition of information of an object or phenomenon, by the use of either recording or real-time sensing device(s) that is not in physical or intimate contact with the object (such as by way of aircraft, spacecraft, satellite, buoy, or ship). In practice, remote sensing is the stand-off collection through the use of a variety of devices for gathering information on a given object or area which gives more information than sensors at individual sites might convey. Thus, Earth observation or weather satellite collection platforms, ocean and atmospheric observing weather buoy platforms, monitoring of a pregnancy via ultrasound, Magnetic Resonance Imaging (MRI), Positron Emission Tomography (PET), and space probes are all examples of remote sensing. In modern usage, the term generally refers to the use of imaging sensor technologies including but not limited to the use of instruments aboard aircraft and spacecraft, and is distinct from other imaging-related fields such as medical imaging.

There are two kinds of remote sensing. Passive sensors detect natural radiation that is emitted or reflected by the object or surrounding area being observed. Reflected sunlight is the most common source of radiation measured by passive sensors. Examples of passive remote sensors include film photography, infra-red, charge-coupled devices, and radiometers. Active collection, on the other hand, emits energy in order to scan objects and areas whereupon a sensor then detects and measures the radiation that is reflected or backscattered from the target. radar, lidar, and SODAR are examples of active remote sensing techniques used in atmospheric physics where the time delay between emission and return is measured, establishing the location, height, speed and direction of an object.

Remote sensing makes it possible to collect data on dangerous or inaccessible areas. Remote sensing applications include monitoring deforestation in areas such as the Amazon Basin, the effects of climate change on glaciers and Arctic and Antarctic regions, and depth sounding of coastal and ocean depths. Military collection during the cold war made use of stand-off collection of data about dangerous border areas. Remote sensing also replaces costly and slow data collection on the ground, ensuring in the process that areas or objects are not disturbed.

Orbital platforms collect and transmit data from different parts of the electromagnetic spectrum, which in conjunction with larger scale aerial or ground-based sensing and analysis, provides researchers with enough information to monitor trends such as El Niño and other natural long and short term phenomena. Other uses include different areas of the earth sciences such as natural resource management, agricultural fields such as land usage and conservation, and national security and overhead, ground-based and stand-off collection on border areas.

Radiation

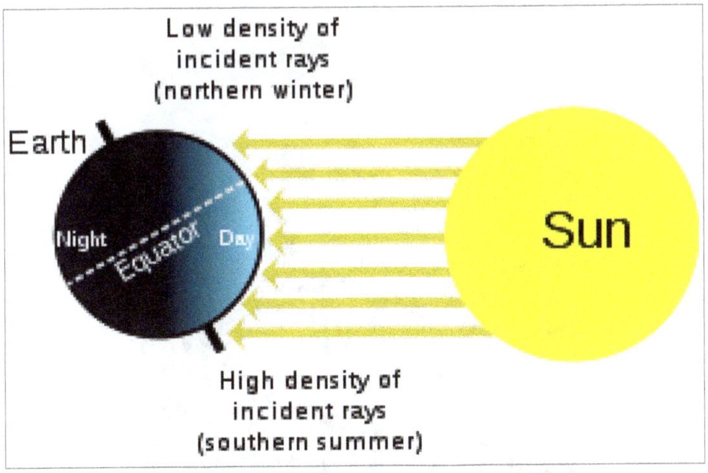

This is a diagram of the seasons. In addition to the density of incident light, the dissipation of light in the atmosphere is greater when it falls at a shallow angle.

Atmospheric physicists typically divide radiation into solar radiation (emitted by the sun) and terrestrial radiation (emitted by Earth's surface and atmosphere).

Solar radiation contains variety of wavelengths. Visible light has wavelengths between 0.4 and

0.7 micrometers. Shorter wavelengths are known as the ultraviolet (UV) part of the spectrum, while longer wavelengths are grouped into the infrared portion of the spectrum. Ozone is most effective in absorbing radiation around 0.25 micrometers, where UV-c rays lie in the spectrum. This increases the temperature of the nearby stratosphere. Snow reflects 88% of UV rays, while sand reflects 12%, and water reflects only 4% of incoming UV radiation. The more glancing the angle is between the atmosphere and the sun's rays, the more likely that energy will be reflected or absorbed by the atmosphere.

Terrestrial radiation is emitted at much longer wavelengths than solar radiation. This is because Earth is much colder than the sun. Radiation is emitted by Earth across a range of wavelengths, as formalized in Planck's law. The wavelength of maximum energy is around 10 micrometers.

Cloud Physics

Cloud physics is the study of the physical processes that lead to the formation, growth and precipitation of clouds. Clouds are composed of microscopic droplets of water (warm clouds), tiny crystals of ice, or both (mixed phase clouds). Under suitable conditions, the droplets combine to form precipitation, where they may fall to the earth. The precise mechanics of how a cloud forms and grows is not completely understood, but scientists have developed theories explaining the structure of clouds by studying the microphysics of individual droplets. Advances in radar and satellite technology have also allowed the precise study of clouds on a large scale.

Atmospheric Electricity

Cloud to ground Lightning in the global atmospheric electrical circuit.

Atmospheric electricity is the regular diurnal variations of the Earth's atmospheric electromagnetic network (or, more broadly, any planet's electrical system in its layer of gases). The Earth's surface, the ionosphere, and the atmosphere is known as the global atmospheric electrical circuit.

Lightning discharges 30,000 amperes, at up to 100 million volts, and emits light, radio waves, x-rays and even gamma rays. Plasma temperatures in lightning can approach 28,000 kelvins and electron densities may exceed $1024/m^3$.

Atmospheric Tide

The largest-amplitude atmospheric tides are mostly generated in the troposphere and stratosphere when the atmosphere is periodically heated as water vapour and ozone absorb solar radiation during the day. The tides generated are then able to propagate away from these source regions and ascend into the mesosphere and thermosphere. Atmospheric tides can be measured as regular fluctuations in wind, temperature, density and pressure. Although atmospheric tides share much in common with ocean tides they have two key distinguishing features:

i) Atmospheric tides are primarily excited by the Sun's heating of the atmosphere whereas ocean tides are primarily excited by the Moon's gravitational field. This means that most atmospheric tides have periods of oscillation related to the 24-hour length of the solar day whereas ocean tides have longer periods of oscillation related to the lunar day (time between successive lunar transits) of about 24 hours 51 minutes.

ii) Atmospheric tides propagate in an atmosphere where density varies significantly with height. A consequence of this is that their amplitudes naturally increase exponentially as the tide ascends into progressively more rarefied regions of the atmosphere. In contrast, the density of the oceans varies only slightly with depth and so there the tides do not necessarily vary in amplitude with depth.

Note that although solar heating is responsible for the largest-amplitude atmospheric tides, the gravitational fields of the Sun and Moon also raise tides in the atmosphere, with the lunar gravitational atmospheric tidal effect being significantly greater than its solar counterpart.

At ground level, atmospheric tides can be detected as regular but small oscillations in surface pressure with periods of 24 and 12 hours. Daily pressure maxima occur at 10 a.m. and 10 p.m. local time, while minima occur at 4 a.m. and 4 p.m. local time. The absolute maximum occurs at 10 a.m. while the absolute minimum occurs at 4 p.m. However, at greater heights the amplitudes of the tides can become very large. In the mesosphere (heights of ~ 50 – 100 km) atmospheric tides can reach amplitudes of more than 50 m/s and are often the most significant part of the motion of the atmosphere.

Aeronomy

Aeronomy is the science of the upper region of the atmosphere, where dissociation and ionization are important. The term aeronomy was introduced by Sydney Chapman in 1960. Today, the term also includes the science of the corresponding regions of the atmospheres of other planets. Research in aeronomy requires access to balloons, satellites, and sounding rockets which provide valuable data about this region of the atmosphere. Atmospheric tides play an important role in interacting with both the lower and upper atmosphere. Amongst the phenomena studied are upper-atmospheric lightning discharges, such as luminous events called red sprites, sprite halos, blue jets, and elves.

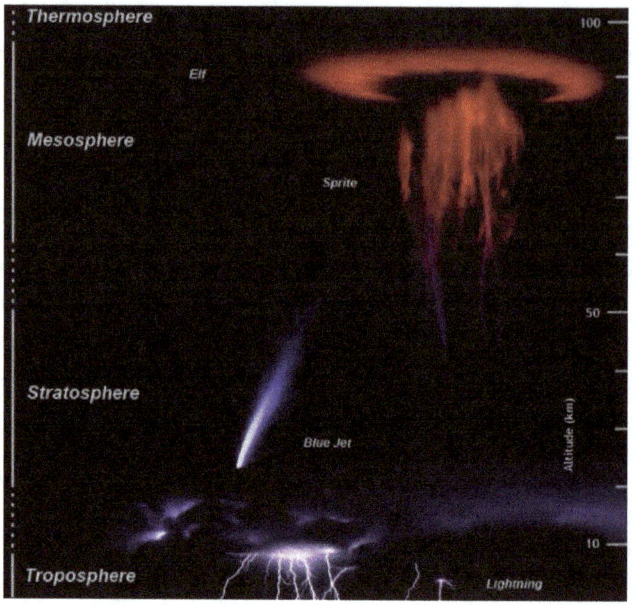

Representation of upper-atmospheric lightning and electrical-discharge phenomena.

Centers of Research

In the UK, atmospheric studies are underpinned by the Met Office, the Natural Environment Research Council and the Science and Technology Facilities Council. Divisions of the U.S. National Oceanic and Atmospheric Administration (NOAA) oversee research projects and weather modeling involving atmospheric physics. The US National Astronomy and Ionosphere Center also carries out studies of the high atmosphere. In Belgium, the Belgian Institute for Space Aeronomy studies the atmosphere and outer space.

References

- Bozorgnia, Yousef; Bertero, Vitelmo V. (2004). Earthquake Engineering: From Engineering Seismology to Performance-Based Engineering. CRC Press. ISBN 978-0-8493-1439-1
- Reinhardt, Carsten (2008). Chemical Sciences in the 20th Century: Bridging Boundaries. John Wiley & Sons. p. 161. ISBN 3-527-30271-9
- Poirier, Jean-Paul (2000). Introduction to the Physics of the Earth's Interior. Cambridge Topics in Mineral Physics & Chemistry. Cambridge University Press. ISBN 0-521-66313-X
- Eratosthenes (2010). Eratosthenes' "Geography". Fragments collected and translated, with commentary and additional material by Duane W. Roller. Princeton University Press. ISBN 978-0-691-14267-8
- Davies, Geoffrey F. (2001). Dynamic Earth: Plates, Plumes and Mantle Convection. Cambridge University Press. ISBN 0-521-59067-1
- Ehrlich, Henry Lutz; Newman, Dianne K.; Kappler, Andreas (2015-10-15). Ehrlich's Geomicrobiology, Sixth Edition. CRC Press. ISBN 9781466592414
- Dilek, Yildirim; Harald Furnes; Karlis Muehlenbachs (2008). Links Between Geological Processes, Microbial Activities & Evolution of Life. Springer. p. v. ISBN 1-4020-8305-X
- Knoll, Andrew H. (2015-03-22). Life on a Young Planet: The First Three Billion Years of Evolution on Earth. Princeton University Press. ISBN 9781400866045

- Bekker, Andrey (2014-01-01). Amils, Ricardo; Gargaud, Muriel; Quintanilla, José Cernicharo; Cleaves, Henderson James; Irvine, William M.; Pinti, Daniele; Viso, Michel, eds. Great Oxygenation Event. Springer Berlin Heidelberg. pp. 1–9. doi:10.1007/978-3-642-27833-4_1752-4. ISBN 9783642278334
- Knoll, Andrew H.; Canfield, Professor Don E.; Konhauser, Kurt O. (2012-03-30). Fundamentals of Geobiology. John Wiley & Sons. ISBN 9781118280881
- L, Slonczewski, Joan; W, Foster, John (2013-10-01). Microbiology: An Evolving Science: Third International Student Edition. W. W. Norton & Company. ISBN 9780393923216

Permissions

All chapters in this book are published with permission under the Creative Commons Attribution Share Alike License or equivalent. Every chapter published in this book has been scrutinized by our experts. Their significance has been extensively debated. The topics covered herein carry significant information for a comprehensive understanding. They may even be implemented as practical applications or may be referred to as a beginning point for further studies.

We would like to thank the editorial team for lending their expertise to make the book truly unique. They have played a crucial role in the development of this book. Without their invaluable contributions this book wouldn't have been possible. They have made vital efforts to compile up to date information on the varied aspects of this subject to make this book a valuable addition to the collection of many professionals and students.

This book was conceptualized with the vision of imparting up-to-date and integrated information in this field. To ensure the same, a matchless editorial board was set up. Every individual on the board went through rigorous rounds of assessment to prove their worth. After which they invested a large part of their time researching and compiling the most relevant data for our readers.

The editorial board has been involved in producing this book since its inception. They have spent rigorous hours researching and exploring the diverse topics which have resulted in the successful publishing of this book. They have passed on their knowledge of decades through this book. To expedite this challenging task, the publisher supported the team at every step. A small team of assistant editors was also appointed to further simplify the editing procedure and attain best results for the readers.

Apart from the editorial board, the designing team has also invested a significant amount of their time in understanding the subject and creating the most relevant covers. They scrutinized every image to scout for the most suitable representation of the subject and create an appropriate cover for the book.

The publishing team has been an ardent support to the editorial, designing and production team. Their endless efforts to recruit the best for this project, has resulted in the accomplishment of this book. They are a veteran in the field of academics and their pool of knowledge is as vast as their experience in printing. Their expertise and guidance has proved useful at every step. Their uncompromising quality standards have made this book an exceptional effort. Their encouragement from time to time has been an inspiration for everyone.

The publisher and the editorial board hope that this book will prove to be a valuable piece of knowledge for students, practitioners and scholars across the globe.

Index

www.ingramcontent.com/pod-product-compliance
Lightning Source LLC
LaVergne TN
LVHW080738210926
848485LV00003B/578